Growth, Poverty and Developmental Aspects of Agriculture

Growth, Poverty and Developmental Aspects of Agriculture: Countries' Perspectives

EDITED BY

RAMESH CHANDRA DAS
Vidyasagar University, India

emerald PUBLISHING

United Kingdom – North America – Japan – India – Malaysia – China

Emerald Publishing Limited
Emerald Publishing, Floor 5, Northspring, 21-23 Wellington Street, Leeds LS1 4DL

First edition 2025

Reprints and permissions service
Contact: www.copyright.com

British Library Cataloguing in Publication Data
A catalogue record for this book is available from the British Library

ISBN: 978-1-83608-077-0 (Print)
ISBN: 978-1-83608-076-3 (Online)
ISBN: 978-1-83608-078-7 (Epub)

Printed and bound by CPI Group (UK) Ltd, Croydon, CR0 4YY

In memory of my colleagues
Dr Kamal Ray and Sanat Kumar Chakraborty

Contents

List of Abbreviations

Chapter 1

ARDL	Auto Regressive Distributed Lag
DEA	Data Envelopment Analysis
FAO	Food and Agriculture Organization
FDI	Foreign Direct Investment
ICU	India, China and United States of America
NGO	Non-Government Organization
TMEC	Tratado entre México, Estados Unidos y Canadá
USCM	United States-Canada-Mexico

Chapter 2

ATMA	National Mission on Agriculture Extension
DAM	Digital Agriculture Mission
DEA	Data Envelopment Analysis
DMU	Decision Making Unit
EMDE	Emerging Market and Developing Economy
FPO	Farmer Producer Organizations
GATT	General Agreement on Tariffs and Trade
MSP	Minimum Support Price
NABARD	National Bank for Agriculture and Rural Development
PMFBY	Pradhan Mantri Fasal Bima Yojana
PMKISAN	Pradhan Mantri Kisan Samman Nidhi
RWBCIS	Restructure Weather Based Crop Insurance Schemes
SFA	Stochastic Frontier Analysis
TE	Technical Efficiency
WTO	World Trade Organization

Chapter 3

ABP	Anchor Borrowers' Programme
ACGSF	Agricultural Credit Guarantee Scheme Fund
AGDP	Agricultural Gross Domestic Product
ALAB	Agricultural Labour
ALAN	Agricultural Land Area
AMACH	Agricultural Machinery
ARDL	Auto Regressive Distributed Lag
BF	Bank Financing
CACS	Commercial Agriculture Credit Scheme
CBA	Commercial Bank's loan to the Agricultural sector
ECM	Error Correction Term
FDI	Foreign Direct Investment
FMARD	Federal Ministry of Agriculture and Rural Development
FSD	Financial Sector Development
GDP	Gross Domestic Product
INT	Interest Rate
NIPC	Nigerian Investment Promotion Commission

Chapter 4

ARDL	Auto Regressive Distributed Lag
BP	British Petroleum
CO2	Carbon Dioxide
CO2E	Carbon Dioxide Emission
CUSUM	Cumulative Sum
ECM	Error Correction Model
EKC	Environmental Kuznets Curve
FAO	Food and Agriculture Organization
FC	Forest Coverage
FDI	Foreign Direct Investment
FDI_AGR	FDI Inflows in Agriculture Sector
GHG	Greenhouse Gas
GtCO2	Gigatonnes Carbon Dioxide
ICU	India, China and United States of America

IPCC	Intergovernmental Panel on Climate Change
Ln	Natural Log
MEA	Millennium Ecosystem Assessment
OECD	Organization for Economic Cooperation and Development
REC	Renewable Energy Consumption
SDG	Sustainable Development Goals
TEEB	The Economics of Ecosystems and Biodiversity
UN	United Nations
USA	United States of America
US	United States
VECM	Vector Error Correction Model
VEC	Vector Error Correction
VA_AGR	Value Added in Agriculture Sector
WPP	World Population Prospects

Chapter 5

AMEG	Mexican Association of Livestock Fattening
GDP	Gross Domestic Product
NAFTA	North American Free Trade Agreement
TMEC	Tratado entre México, Estados Unidos y Canadá
USCM	United States-Canada-Mexico

Chapter 6

AGROSE GURO	Agrupación Española de Entidades Aseguradoras de los Seguros Agrarios Combinados, Spain
AUSAID	Australian Agency for International Development
AUTP	Asuransi Usaha Tani Padi, Indonesia
CCIS	Comprehensive Crop Insurance Scheme
CNY	Chinese Yuan Renminbi
CPIS	Coconut Palm Insurance Scheme
ECIS	Experimental Crop Insurance Scheme
FIIS	Farm Income Insurance Scheme
FNGCA	National Agricultural Disaster Fund, France
JICA	Japan International Cooperation Agency

MNAIS	Modified National Agricultural Insurance Scheme
MPCI	Multi-Peril Crop Insurance
NAIS	National Agricultural Insurance Scheme
NBS	Nature Based Solutions
NCIP	National Crop Insurance Programme
NDVI	Normalized Difference Vegetation Index
NGO	Non-Government Organization
NTB	Negusa Tengara Barat, Indonesia
PCIS	Pilot Crop Insurance Scheme
PICC	Peoples Insurance Company of China
PMFBY	Pradhan Mantri Fasal Bima Yojana
PPP	Public–Private Partnerships
PROAGRO	Programa de Garantia da Actividade Agropequária, Guarantee Programme for Agricultural Activities, Brazil
PRONAF	Programa Nacional de Fortalecimento da Agricultura Familiar, Brazil
PSSCI	Pilot Scheme on Seed Crop Insurance
SEAF	Seguro da Agricultura Familiar, Insurance for Family Agriculture, Brazil
USD	United States Dollar
UT	Union Territory
WBCIS	Wather Based Crop Insurance Scheme

Chapter 7

AIC	Akaike (1969) Information Criterion
GARCH	Generalized Autoregressive Conditional Heteroscedasticity
MGARCH	Multivariate Generalized Autoregressive Conditional Heteroscedasticity
ARCH	Autoregressive Conditional Heteroscedasticity
MANOVA	Multivariate Analysis of Variance
RML	Reuters Market Light

Chapter 8

AIBP	Accelerated Irrigation Benefit Programme
ANOVA	Analysis of Variance

NPK	Nitrogen, Phosphorus, and Potassium
RBI	Reserve Bank of India
TFPG	Total Factor Productivity Growth

Chapter 9

CDMP	Comprehensive Disaster Management Programme
CPP	Cyclone Preparedness Programme
DMB	Disaster Management Bureau
DM	Disaster Management
DRR	Disaster Risk Reduction
ERM	Emergency Response Management
FAP	Flood Action Plan
FGD	Focus Group Discussion
GR	Gratuitous Relief
LGIs	Local Government Institutions
MoDMR	Ministry of Disaster Management and Relief
NDMC	National Disaster Management Council
NGO	Non-Governmental Organization
NPDM	National Plan for Disaster Management
OMS	Open Market Sale
SOD	Standing Orders on Disaster
SSNP	Social Safety Net Programs
SPSS	Statistical Package for the Social Sciences
UP	Union Council
VGD	Vulnerable Group Development
VGF	Vulnerable Group Feeding

Chapter 10

ADF	Augmented Dickey-Fuller
ARDL	Autoregressive Distributed Lag
ECT	Error Correction Term
FAO	Food and Agriculture Organization
OECD	Organisation for Economic Co-operation and Development
PP	Phillips-Perron

VAR	Vector Autoregressive
VECM	Vector Error Correction Model

Chapter 11

ASHA	Accredited Social Health Activist
CAGR	Compound Annual Growth Rate
CSAPs	Climate-Smart Agricultural Practices
DAC & FW	Department of Agriculture, Cooperation and Farmers Welfare
DES	Directorate of Economics and Statistics
FAO	Food and Agriculture Organization
GDP	Gross Domestic Product
GHI	Global Hunger Index
GVA	Gross Value Added
ICDS	Integrated Child Development Services
Kg	Kilograms
MGNREGA	Mahatma Gandhi National Rural Employment Guarantee Act
MT	Million Tonnes
NABARD	National Bank for Agriculture and Rural Development
NFSA	National Food Security Act
PDS	Public Distribution System
PMGKY	Pradhan Mantri Garib Kalyan Yojana
PPP	Public Private Partnership
R&D	Research & Development
RFID	Radio Frequency Identification
SEZ	Special Economic Zones
TISM	Total Interpretive Structural Modelling
TPDS	Targeted Public Distribution System
WFP	World Food Programme
WFS	World Food Summit

Chapter 13

ACC	Accessibility
ADF	Augmented Dickey-Fuller

AFF	Affordability
ARDL	Autoregressive Distributed Lag Model
AV	Availability
AWW	Agricultural Wage of Women Worker
CI	Cost of Education
CV	Coefficient of Variation
DFE	Difference Fixed Effect
ECM	Error Correction Model
EPWRF	Economic and Political Weekly Research Foundation
FAO	Food and Agricultural Organization
IBRD	International Bank for Reconstruction and Development
KMO	Kaiser–Meyer–Olkin
MDFSI	Multidimensional Food Security Index
MG	Mean Group
NFHS	National Family Health Survey
NIA	Net Irrigated Area
NSDP	Net State Domestic Product
NSSO	National Sample Survey Organization
PCA	Principal Component Analysis
PCE	Per Capita Electric Consumption
PMG	Pooled Mean Group
RBI	Reserve Bank of India
SDG	Sustainable Development Goals
STA	Stability
TLR	Total Length of the Road
UNDP	United Nations Development Programme
WFPR	Workforce Participation Rate

Chapter 14

ADB	Asian Development Bank
BHA	Bureau for Humanitarian Assistance
UNDP	United Nation Development Programme
USAID	United States Agency for International Development

Chapter 15

SDG	Sustainable Development Goal
GSDP	Gross State Domestic Product
FINP	Food Insecure but Non-Poor
FSP	Food Secure but Poor
FIP	Food Insecure and Poor
FSNP	Both Food Secure & Non-Poor
MPI	Multidimensional Poverty Index
OLS	Ordinary Least Squares
HCR	Head Count Ratio

About the Editor

Ramesh Chandra Das, PhD, is currently a Professor and the Head of the Department of Economics, Vidyasagar University, India, with 25 years of teaching and research experience. Professor Das has published more than 100 articles in reputed journals and has written and edited more than 20 books. His research interests lie in applied macroeconomics, public economics and environmental economics.

About the Contributors

Soumyajit Adhikari is currently pursuing the degree Master's of Arts in Economics from University of Calcutta, Kolkata, West Bengal, India. He has certain publications from well reputed journals and books. His areas of interest are international trade, resource and environmental economics, microeconomic modelling, labour economics and open economy macroeconomics.

Olumide Abiodun Ayetigbo is a Lecturer at the Department of Business Administration, Faculty of Management Sciences, National Open University of Nigeria, Jabi, Abuja, Nigeria. He holds a Master in Business Administration (MBA Finance) and a Master of Science (MSc) in Business Administration, respectively, from the prestigious Obafemi Awolowo University, Osun State, Nigeria.

Saheli Bhattacharya, PhD, after completing MCom from University of Calcutta and qualifying UGC-NET, she has been attached with various reputed colleges of Kolkata including Sivanath Sastri College (as a full-time Contractual Lecturer) and Seth Anandaram Jaipuria College (as a Guest Lecturer). She has co-authored several articles which have been published in reputed journals and also presented papers in international conferences. She completed her PhD from University of Calcutta in 2023.

Somdip Bhukta is currently a doctoral-level Research Scholar at the Department of Economics, Vidyasagar University, India. He has been awarded the Junior Research Fellowship by the University Grants Commission, India. His research interest lies in the areas of agricultural economics, environmental economics and development economics.

Bappaditya Biswas, PhD, is currently working as an Assistant Professor in the Department of Commerce, University of Calcutta, India, with over 15 years of teaching experience in undergraduate as well as postgraduate courses. Dr Biswas has published more than 35 articles in different journals and edited a volume of repute.

Nilendu Chatterjee, PhD, is an Assistant Professor in the Department of Economics, Bankim Sardar College, West Bengal, India. He has research interest in resource economics, general equilibrium and development economics. He has published a number of research articles in several international journals of economics including *International Journal of Sustainable Economies Management*, *Economic Affairs* and *Foreign Trade Review*.

Arup Kumar Chattopadhyay, PhD, is a retired Professor of Economics at the University of Burdwan, India, with more than 30 years of teaching experience. His research articles have been published in numerous national and international peer-reviewed journals in the areas of financial economics, agricultural economics, public finance, food economics, operations research and econometrics. He was the Dean of Faculty Council for PG Studies in Arts, Commerce, Law, Music and Fine Arts at the University of Burdwan.

Arpita Das is a PhD Research Scholar at IIT Guwahati, India. Her interest area lies in natural resource economics, environmental economics and development economies. She holds both MA and MPhil degrees in Economics from Utkal University, India. She is also focusing on GIS application in the field of Economics.

Sunanda Das has completed her Master's in Economics from Vidyasagar University, India, with first class. She also holds a diploma in Data Science from the same university. She has research interests on agricultural economics, social sector, inclusive growth and development.

Tulshi Kumar Das, PhD, is a Professor of Social Work at Shahjalal University of Science and Technology, Sylhet, Bangladesh. He obtained his PhD and MPhil degrees from Jawaharlal Nehru University, New Delhi, India. He has widely published articles with internationally reputed publishers. He has successfully completed many research projects relating to domestic violence, indigenous and international social work, social services and COVID-19, natural disasters, food security, ethnicity, gender inequality, human rights, nationalism, and genocide sponsored by Comic Relief Fund, UK, FAO, Ministry of Social Welfare, and European Commission, among others.

Hasan Dincer, PhD, is an academician in the interdisciplinary fields of business and finance, holding the position of Professor at Istanbul Medipol University in Turkey. With over 400 scientific articles to his name, many of which are indexed in high-impact journals like SSCI, SCI-Expanded and Scopus, Prof Dincer has contributed significantly to the academic literature. He has focused his research work on a range of topics such as strategic management, innovation, sustainability, circular economy, decision making as well as renewable energy investments, finance and banking.

Jennifer Fernandes, PhD, is currently working as an Associate Professor in the Department of Economics, School of Humanities and Social Sciences, JAIN (Deemed-to-be University), Bangalore. She was awarded a PhD in Economics from Bangalore University, Bangalore.

Dr Tanushree Gupta is currently working as an Assistant Professor in the Faculty of Business Management and Commerce, JSPM University, Pune, Maharashtra. She has 8 years of experience in the field of academics and research. She is an imminent researcher. She earned her Master's in Business Administration from Rajasthan Technical University, Kota. She later received her PhD from Amity University Rajasthan, Jaipur. Finance, microfinance, stock market, and international business are some of her primary academic interests. She is a lifetime member of the All India Commerce Association. She is also a review member of the *International Journal of Research and Analytical Reviews* (Peer peer-reviewed, Referred Journal) and a member of an editorial board of the *International Journal in Management and Social Science* (international peer-reviewed journal). Last but not least, she has in her credit many research papers/publications of national and international repute, and many of her book chapters have been published in international and Scopus-indexed journals.

Md Mahmud Hasan has been working as an Assistant Professor in the Department of Public Administration at Shahjalal University of Science and Technology (SUST) in Sylhet, Bangladesh, since 2012. Before joining SUST, he made a significant contribution to the two prominent research institutions in Bangladesh: the BRAC Development Institute (BDI) at BRAC University and the Power and Participation Research Center (PPRC). His research interests include the impacts of Western imperialist policies in South Asia, education culture, climate change, urban poverty, and local government.

Sudipta Jha, PhD, has been an Assistant Professor in the Department of Economics of Vidyasagar University since April 2012. She obtained her PhD degree from Jadavpur University in the year 2017. Currently, areas of her interest are macroeconomic theory and international macroeconomics with a special interest in Indian macroeconomy.

Susobhan Maiti, PhD, is an Assistant Professor in the Department of Economics, JAIN (Deemed-to-be University), Bangalore, India. He has published many research papers in the areas of industry, macroeconomics and development economics in both national and international journals.

Prasenjit Makur is an Assistant Professor of Commerce at Calcutta Girls' College, Kolkata, and West Bengal. He obtained his graduation and Master's degrees from the University of Burdwan, India. He has more than 7 years of teaching experience in the fields of accounting, finance, taxation and economics and more than 6 years of research experience, with numerous publications in peer-reviewed national journals and books.

Syed Sikander Mehdi, PhD, educated in Dhaka University and Australian National University, is a retired Professor and a leading Peace Scholar from Pakistan. Besides teaching peace studies in the universities in Pakistan, he offered courses on peace studies in universities in Austria and Spain. He also availed research fellowships in Norway, USA and Japan. He has published extensively on wars and war victims, nuclear insecurity, Islam and nonviolence, forced migration issues, culture of peace and museums for peace.

Saddam Hossain Mondal, PhD, is an Assistant Professor of Economics in the Department of Economics, Heramba Chandra College under Calcutta University, India. He did his doctorate degree at the University of Burdwan on Patterns of Foodgrains Production and Food Security in India. His areas of interest include macroeconomics, agricultural economics, food economics and environmental economics.

Partha Mukhopadhyay, PhD, has been a Research Scholar in the Department of Humanities and Social Sciences, National Institute of Institute Durgapur, India, and works at the Business Excellence Department of Duragpur Steel Plant, Steel Authority of India Limited. His research interests include agricultural management, business management, and applied econometrics. He has a number of publications in reputed journals and books.

Debasish Nandy, PhD, is an Associate Professor and Head, Department of Political Science, Kazi Nazrul University, Asansol, West Bengal, India. Dr Nandy is a Visiting Faculty in the Department of Foreign Area Studies, Tajik National University, Republic of Tajikistan. His research areas include India's foreign policy, South Asian politics, South-East Asian affairs and security studies.

Biswajit Nayak is a PhD Research Scholar in the Department of Economics of Vidyasagar University, India. Currently, areas of his research are under the broad head of macroeconomics where he specifically studies economic development, government expenditure and the efficacy of fiscal policy.

Olalekan Charles Okunlola, PhD, is an Associate Lecturer at the University of Derby and Leeds Trinity University in the United Kingdom. He has also held the position of Senior Research Fellow at the Institute for Peace and Conflict Resolution in Nigeria, as well as serving as an Adjunct Lecturer at the National Open University of Nigeria. With a PhD in Economics, his research interests encompass econometrics, development economics, institutional economics and defence economics.

Elsa Patricia Orozco Quijano, PhD, is an Associate Professor at Laurentian University, Canada, with a long career in teaching and research in subjects like international trade, institutional capital, social capital, organizational theories, among others with a list of good publications in internationally reputed scientific journals.

Bidur Paria, PhD, an Assistant Professor of Economics at Dr B R Ambedkar School of Economics University Bengaluru, India, earned his PhD from the Indian Institute of Technology Kharagpur. His academic prowess extends to his significant contributions published in esteemed international journals, such as *Applied Sciences*, *World Development Perspectives* and *ADBI Working Paper* from the Asian Development Bank Institute.

Durlav Sarkar, PhD, is currently working as the Registrar of Raiganj University, West Bengal. Formerly, Dr Sarkar served as an Associate Professor in Management at Royal University of Bhutan, and prior to that, he worked as a Professor and Director at Delhi School of Professional Studies and Research, New Delhi. With more than 20 years of teaching, research, consultancy, administrative and industrial experiences, he has the credits of various publications mainly in the area of human resource management.

Madhabendra Sinha, PhD, is an Assistant Professor at Department of Economics and Politics, Visva-Bharati University, Santiniketan, India, and previously he served as an Assistant Professor at Department of Business Administration, Raiganj University, India. His research interests include development economics, international economics and applied econometrics, with a number of SSCI/Scopus/ABDC indexed publications from Elsevier, Springer, Routledge, Wiley, Sage, Emerald, World Scientific, and Inderscience.

M. C. Omar C. Vargas-González, PhD, is a Professor and the Head of the Department of Systems and Computing at the Technological National of Mexico Campus Ciudad Guzmán, México. His areas of research interests lie in business administration, entrepreneurial economics, computing in social science, arts and humanities and mathematics education.

José G. Vargas-Hernández, PhD, is currently a Research Professor, Instituto Tecnológico Mario Molina Pasquel y Henríquez, Unidad Académica Zapopan, Zapopan, México. He has a long career in Administrative and Management Sciences having more than 200 articles published in different journals, books, and reports.

Foreword

In an era defined by rapid technological advancements and shifting global dynamics, the importance of agriculture remains undiminished. It is both the foundation of economies and the frontline in the battle against poverty and food insecurity. This makes *Growth, Poverty and Developmental Aspects of Agriculture: Countries' Perspectives*, edited by Professor Ramesh Chandra Das, a timely contribution to the ongoing dialogue on how agriculture can continue to fuel development, sustain populations and alleviate poverty across various regions, particularly in developing and emerging countries. Professor Das, along with a distinguished panel of contributors, provides a compendium of research that navigates through the multifaceted role of agriculture in economic growth and development. The book is structured in two insightful sections, with the first focusing on the growth dynamics of agriculture and the latter delving into the intertwined issues of poverty, food security and development.

The relevance of this book cannot be overstated. At a time when global food security faces threats from climate change, political instability and economic disparities, the studies contained within offer crucial insights. The research not only assesses current agricultural practices and their efficiencies, such as in the production of food grains across Indian states and the sustainability of agricultural growth in Pakistan, Bangladesh, Nigeria like countries and G20 Groups, but also addresses complex global issues like food price volatility and the economic theories surrounding food production and security.

What sets this book apart is its global perspective, capturing experiences from Nigeria, Mexico, Pakistan and India, among others, along with several groups of economies. Each chapter contributes to a richer understanding of the local challenges and solutions that define agricultural growth and its impact on poverty reduction and food security. For instance, the investigation of agricultural exports under renegotiated trade agreements offers a keen look at how international policies influence local agricultural practices and outcomes. The timing of this publication is particularly significant. As nations grapple with the Sustainable Development Goals (SDGs), understanding the link between agricultural practices and poverty alleviation is critical. This book not only contributes to academic and policymaking discourses but also serves as a vital resource for stakeholders aiming to forge a path towards sustainable agricultural practices that support economic growth and ensure food security. Emerald Publishing should be commended for bringing such a crucial piece of scholarly work to the public, aiding in the dissemination of knowledge that could very well dictate the quality of life and economic stability for billions of people worldwide.

Hasan Dincer

Preface

It is still verifiable that the agricultural sector has been the vivacious decider within the progression of a country, indeed within the period of technological advancements. With the presentation of innovative headways within the sector, most of the nations from the south are currently experiencing greater well-being, less poverty and destitution and food security. The impacts got boosted when the universal exchange perspective of the division was taken into consideration. Assisted by the development of agrarian yield, numerous nations of the world have made it conceivable to diminish the wage crevices between the less created and the created nations through wage joining. Having a critical share in add up to yield of an economy, its instability in production, abdicate rates, costs, etc. may moreover cause the same for the full economy. Consequently, solidness issue along the related agro-variables gets to be one of the critical plans of the present research attempt.

Under the milieu, the present edited book titled *Growth, Poverty and Developmental Aspects of Agriculture: Countries' Perspectives* throws light upon compiling studies related to the growth, stability, crop insurance, poverty and food security aspects of the agricultural growth in the countries and groups, especially from the developing world.

The book is compiled of 15 chapters which are divided into two broad sections. Section I covers the growth aspects of agriculture across the countries and groups, containing eight chapters including this introductory chapter, and Section II captures the studies related to the poverty, food security and developmental aspects of agriculture across the countries and groups covering seven remaining chapters.

The summary observations are that the countries have improved their growth in the agricultural sector leading to food insecurity and poverty reductions that helped in the overall development of the citizens. The institutional factors such as credit, insurance, irrigation and good governance have instigated the growth process of the sector. Besides the glorious achievements, the sector has faced fluctuations and instability in its production and productivity in several crops at the country as well as regional levels, which lead to hurdles in achieving sustainability in the sector. In addition, in some regions of India, it was observed that the non-poor section of the society is relatively less food-secured compared to the poor section of the society, making the social imbalance in the reverse directions. Thus, the overall impression from the book is that the governments of the countries, states and local regions should come forward with proper governance and institutional facilities to overcome the sustainability problems arising out of the agriculture sector.

The contents of the book may provide thought-provoking solutions to the existing problems of growth, poverty, food security and overall developmental issues of the agriculture sector, and there may be the possibilities of value addition to the existing literature in this area. Readers and policymakers will benefit immensely from the contents of the book.

While carrying out the book project, the editor got tremendous support from different corners of the academic and social world. The editor acknowledges the support of the entire Emerald Team, the contributing authors and the Foreword writer. In addition, the editor is indebted to his parents, wife, daughter and other members of the family for their sacrifice and support to unveil the title. However, the editor is solely responsible for any error still left in the book.

Ramesh Chandra Das

Section I

Growth Aspects of Agriculture Across the Countries and Groups

Chapter 1

Growth, Poverty and Developmental Aspects of Agriculture: An Introductory Note

Ramesh Chandra Das

Vidyasagar University, India

Abstract

From the history of economic development for all types of economies of the world, agriculture has been a major source of income growth, food security and development. It applies traditional as well as modern tools and technologies to maintain rising growth trends, although in some cases it makes the environment polluted. The developing countries of today, those who were suffering from poverty and food insecurity, have now been able to grow several crops and feed their growing population. Productivity has been rising in the sector with the support of institutional interventions, such as credit and financing facilities, crop insurance facilities, irrigation facilities, fertilizer provisions, among others, leading to more production with the same level of two traditional inputs such as labour and capital. The world has been facing, on the other hand, the fluctuations and instabilities in the trends of different types of agricultural productions and yield rates. The present book, with the background, aims to cover up studies on the poverty increase and food security aspects of agriculture in several countries and populations. The summary of the studies shows that there has been growth in agricultural production and productivities and in maintaining more food security in the covered countries with institutional support but with rising fluctuations. The governments of these countries are recommended to intervene and make appropriate institutional support in an equitable manner across their regions so that rising growth could be maintained in a steady path in the agricultural sector.

Growth, Poverty and Developmental Aspects of Agriculture, 3–9

Published under exclusive licence by Emerald Publishing Limited
doi:10.1108/978-1-83608-076-320241001

Keywords: Agricultural sector; growth; poverty; food security; development; institutions

Introduction

It is undeniable that agricultural growth has been the important determiner in the progress of a nation, even in the era of technological advancements. With the introduction of technological advancements in the sector, most of the countries from the south are now experiencing good well-being, less poverty and increased food security. The effects are boosted when the international trade aspect of the sector is taken into consideration. Furthermore, with the growth of agricultural output, many countries across the world have made it possible to reduce the income gap between the less developed and the developed countries through income convergence. Having a significant share in the total output of an economy, its volatility in production, yield rates, prices, etc. may also cause the same for the whole economy. Hence, stability issues for the related agro-variables become one of the important agendas of the present research. From the history of economic development for all types of economies of the world, agriculture has been a major source of income growth, food security and development. It applies traditional as well as modern tools and technologies to maintain rising growth trends, although in some cases it causes environmental pollution. The developing countries of today, those who were suffering from poverty and food insecurity, have now been able to grow several crops and feed their growing population. The productivity has been rising in the sector with the support of institutional interventions such as credit and financing facilities, crop insurance facilities, irrigation facilities, fertilizer provisions, among others, leading to more production with the same level of the two traditional inputs such as labour and capital (Čolović & Petrović, 2014; Das, 2023; Das et al., 2018; Duxbury, 2001; Enjolras et al., 2012; Gangopadhyay, 2004). The world has been facing, on the other hand, the fluctuations and instabilities in the trends of different types of agricultural productions and yield rates.

The problem of food security depends much on food availability along with access, utilization and stability (Clay, 2002; FAO, 2006; Frongillo, 1999). This book will focus on the food availability criterion including both food production and trade in foods. The nature of agriculture is such that every nation is not capable of producing every product with the same quality and with the same efficiency. Therefore, food security in the modern era means not only a rise in per capita production but also a successful engagement in the international trade both as an exporter and an importer. Now, with the rise in food prices, both production and export of food are expected to increase in the domestic as well as in the international market. With the profitability of the agriculture, investments and productivity of the agricultural sector will also rise. Hence, a remunerative farm price seems to be the precondition of food security. In addition, crop insurance schemes have been successful in many countries, leading to food supply guaranteed at the suppliers' end. The existing literature reveals that there has been growth in the agricultural sector worldwide with fluctuations due to various causes. The growth of the sector leads to food security of the world population and low level of poverty, especially of the south, even if there are issues related to agro-products exports, imports, price volatility, etc.

Objectives of the Book

Under the aforementioned backdrop, *Growth, Poverty and Developmental Aspects of Agriculture: Countries' Perspectives* aims to compile studies related to the growth, stability, crop insurance, poverty and food security aspects of the agricultural growth in different countries and populations, especially from the developing world.

Review and Discussion on the Chapters

The book is composed of 15 chapters, which are divided into two broad sections: Section I covers up the growth aspects of agriculture across the countries and populations containing eight chapters including this introductory chapter, and Section II captures the studies related to the poverty, food security and developmental aspects of agriculture across the countries and populations covering the remaining seven chapters. A brief sketch of all the chapters is outlined here.

Chapter 2 explores various aspects of production of food grains across 30 Indian states, and it emphasizes on measuring the efficiency of food grain production of all the 30 states on the basis of non-parametric Data Envelopment Analysis (DEA) with the additional consideration of economies of scale. The results highlight the fact that the Indian states are about 21% inefficient in terms of food grain production with the average efficiency score being 0.79. The methodology adopted for this study incorporates crucial factors such as usage of land area, usage of fertilizers and allocation of bank credit to carry the analysis forward.

Chapter 3 examines the short- and long-term effect of deposit money banks' loan on Nigeria's agricultural sustainability for the period 1985–2017 using the autoregressive distributed lag (ARDL) bonding test. The study found that bank loans have a negative impact on agricultural sustainability in both the short and long terms. Along the same line, deposit money banks' interest rate also has an inverse relationship on Nigeria's agricultural sustainability in both the long and short terms. The study concludes that bank finance is very impactful in stimulating agricultural sustainability within the context of the Nigerian economy.

Chapter 4 investigates the causal association and short- and long-term effects of foreign direct investment (FDI) inflows, forest coverage and CO_2 emissions on the agricultural productivity of the India, China and US (ICU) economies for the period 1990–2023 using the ARDL test. The results confirm that FDI inflows have a significant and positive impact on Indian and Chinese agriculture productivity, whereas CO_2 emissions adversely affect US agriculture productivity in the long term. In the short term, CO_2 emissions led to agricultural productivity in both China and the US economies. The bound test and error correction mechanism (ECM) result also confirmed the long-term connection and convergence of the equilibrium path among the studied variables except India. It is thus recommended to encourage FDI in the agriculture sector and expand the use of environmentally friendly technology to decrease carbon emissions and promote forest coverage for sustainable growth and higher agricultural production.

Chapter 5 deals on Mexico which has invested in finding measures to attract FDI to its territories by signing the United States–Canada–Mexico (USCM) treaty in 1994 that posits it as the largest direct exporter of the United States, who is likely to withdraw from the treaty after the renegotiation of the USCM aims to determine the advantages and disadvantages of doing so. It is contended that the renegotiation of the TMEC does not have to put the agricultural commodities and products from the State of Sinaloa in a wave of uncertainty regarding its exports. The study suggests that such a renegotiation is not beneficial for Sinaloan agricultural exportation putting forth the evidence to support the hypothesis employed.

Chapter 6, citing crop insurance being the protectors to the farmers and helping in reducing any element of chance associated with agriculture, intends to explain the role of crop insurance towards sustainable agriculture and to make a study of the crop insurance schemes prevalent in the selected European and Asian countries. The study concludes that crop insurance is being offered in the selected European and Asian countries in various forms. Public–private partnership is very common among these selected countries.

Chapter 7, quoting uncertainty in commodity pricing as a major cause of concern for farmers of developing countries like India, intends to observe the decision-making process of farmers of Burdwan district of West Bengal, India, using the generalized autoregressive conditional heteroscedasticity (GARCH) model to measure the extent of volatility in spot prices of potato and also to observe the existence of seasonal effects, over the period of 2003–2019. It finds that the market price of potato decreased during its season of production and peaked during the off-season period. The result implies that the volatility of potato forecasting is tending towards the standard error correction in the long term; from 2003 to 2019, the trend of potato price was influenced by multiple events. The result indicates that the farmers must focus on short-term structural events in the potato market.

Chapter 8 focuses first on the stability analysis of quantity of food production and irrigation facilities in the top 10 rice-producing states in India and then on the stability of yield rates in three varieties of paddy production: Aus, Aman and Boro. It also focuses on irrigation facilities across the blocks in the Paschim Medinipur district of West Bengal, as a special case. It studies the growth and stability by means of trend analysis and Coppocks method and finds out that Uttar Pradesh and Andhra Pradesh have a positive significant correlation between irrigation facilities and paddy production. It is also found that the correlation between changes in the yield rate of paddy production and changes in irrigation is positive and significantly related to some agriculture-rich blocks in Paschim Medinipur district under West Bengal. Therefore, the results lead us to prescribe that the government should focus on irrigation facilities in those areas where the effects of irrigation facilities are realized well.

Chapter 9 explores the impact of flash flooding in Bangladesh on food security, employment and the overall well-being of the affected population; it also evaluates government and NGO initiatives taken to alleviate the crisis in Bangladesh by adopting a mixed-methods approach and incorporating qualitative and quantitative data collection techniques through surveys of the five affected villages. It finds out that the flash flood has caused severe damage to the agricultural sector, with almost every farmer in the affected area losing their single-season high-yielding *boro* paddy, resulting in year-long food insecurity and unemployment. The government responded to the situation by providing year-long food and monetary aid through various social safety net programmes.

Chapter 10 analyses the impact of a rising food prices on the problem of food security of the G-20 countries from the perspective of food production. The study has taken rice as a representative food item, and using the ARDL technique; it has found that rice production is significantly responsive to the domestic relative price of rice. However, there is no price incentive operative for rice productivity. Rather, it finds the evidence of a negative relation between rice productivity and domestic rice prices in relation to the global prices.

Chapter 11, pointing out the availability of food, access to food and absorption of food as the three key aspects of food security, intends first to measure the annual growth rate and compound annual growth rate (CAGR) of production of food grains, nutri coarse cereals, total pulses and nine oilseeds for the period 2011–2021 in India, and then, to estimate the per capita availability of these crops as a measure of the food security in India. It observes that the highest production is recorded in between 2011 and 2021 for food grains (315.72), followed by the production of nine oil seeds (36.7), total pulses (27.69) and nutri coarse cereals (26.42). The per capita availability of food grains increased from 206.17 kg in 2011 to 224.30 kg in 2021. The per capita availability of total pulses increased from 13.59 kg in 2011 to 19.67 kg in 2021, which indicates India has managed food security successfully in the past decade.

Chapter 12 explains how a predominantly agricultural country, Pakistan, has been reduced into a food resource poor country, and how has the negligence of social, economic and cultural sectors made Pakistan more insecure. This study mainly looks for an answer to two questions: first, should food insecurity be studied in isolation as a separate issue or as a part of the larger issue of governance, security and people; and second, isn't there a need to drastically restructure the economic, political and security landscape and framework to significantly diminish poverty, illiteracy and food insecurity in the country?

Chapter 13 examines the status of food security for selected states in India during 1995–1996 to 2018–2019 where the statuses of food security for the states are estimated by applying the multidimensional food security index. The study has also used panel ARDL method for the estimation of log-term and short-term relationship between food security and socio-economic indicators of different states. It is observed that the value of the food security index has significantly increased over the study period. The food security conditions for all the states have improved at higher rates in the second sub-period compared to the first sub-period in the country. Furthermore, the food security of most of the states is associated with the per capita power consumption, workforce participation rate and cropping intensity in the country. In addition, the regional imbalance has decreased, establishing that food security of the poorer states improved at a higher rate compared to the richer states.

Chapter 14 intends to investigate the causes and consequences of food security in Pakistan and Afghanistan through the content analysis method. Due to the acute financial crisis since 2021, both countries are searching for a way out of this economic hardship. Some common problems in both countries include lack of industrialization, ethnic conflicts, political violence, terrorism, corruption, mis-governance, arms trafficking, drug trafficking and human trafficking, poverty, illiteracy, over-dependency on foreign aid, etc. The study reveals that due to wrong governmental policy, corruption, lack of agriculture policy and food security policy, food security is a very common basic problem for both countries. For Pakistan, over-expenditure on the military head and the Chinese debt trap made the situation more complicated. Again, due to the shortage of water, modern irrigation systems, droughts and modern agriculture equipment, Afghanistan could not produce adequate food for their people.

The final chapter, Chapter 15, critically reviews the methodologies applied to measure household poverty and food insecurity and examines the relationship between the two applying the m-logit model for data analysis using primary data collected from two villages of the Balasore district of Odisha, whose major livelihood relies on agriculture. The results show that even households living above the poverty line suffer from the problem of food insecurity. About 20.5% of the non-poor are food insecure in the study villages. Hence, there is a need for properly targeted policy interventions and to revise the methodologies applied to measure household poverty and identify beneficiaries so that non-poor households are not deprived of food security programmes.

Conclusion

The studies covered in the book have mainly aimed at examining the poverty increase, food security and developmental aspects of agriculture in countries and populations. The summary observations are that the countries have improved their growth in the agricultural sector leading to food security and poverty reductions that helped in the overall development of the populations. Institutional factors such as credit, insurance, irrigation and good governance have instigated the growth process of the sector. Besides these glorious achievements, the sector has faced fluctuations and instabilities in its production and productivity in several crops at the country as well as regional levels, which leads to difficulty in achieving sustainability in the sector. In addition, in some regions of India, it was observed that the non-poor section of the society has relatively less access to food compared to the poor section of the society, thereby making a reverse social imbalance. Thus, the overall impression from the book is that the governments of the countries, states and local regions should step up with appropriate governance and institutional facilities to overcome the sustainability problems arising out of the agricultural sector.

References

Clay, E. (2002). Food security: Concepts and measurement. *Trade Reforms and Food Security: Conceptualising the Linkages, 2002*, 25–34.

Čolović, V., & Petrović, N. M. (2014). Crop insurance-risks and models of insurance. *Economics of Agriculture, Institute of Agricultural Economics, 61*(3), 561–573.

Das, R. C. (2023). Sustainability through total factor productivity growth in agriculture incorporating institutional factors: A post-globalized Indian scenario. *International Journal of Social Ecology and Sustainable Development, 14*(1), 1–16.

Das, R. C., Das, A., & Ray, K. (2018). Examining forward and backward linkages between public and private investments: A cross-country analysis. *Review of Market Integration, 10*(1), 1–31.

Duxbury, J. M. (2001). Long-term yield trends in the rice-wheat cropping system. *Journal of Crop Production, 3*(2), 27–52.

Enjolras, G., Capitanio, F., & Adinolfi, F. (2012). The demand for crop insurance: Combined approaches for France and Italy. *Agricultural Economics Review, 13*(1), 5–22.

Food Policy Brief. (2006). Food security. *FAO Agriculture and Development Economics Division*, (2).

Frongillo, E. A., Jr. (1999). Validation of measures of food insecurity and hunger. *The Journal of Nutrition, 129*(2), S506–S509.

Gangopadhyay, S. (2004). An alternative to crop insurance. *Economic and Political Weekly, 39*(44), 4763–4764.

Chapter 2

Assessing Efficiency of Food Grain Production Across Indian States: An Investigation Based on Data Envelopment Analysis

Soumyajit Adhikari[a] *and Nilendu Chatterjee*[b]

[a]University of Calcutta, India
[b]Bankim Sardar College, India

Abstract

Over the past decade, India has emerged as one of the major exporters of agricultural products in the international market. Although agriculture in India accounts for about 50% of the economy's employment, its contribution as share to India's gross domestic product is significantly low. India primarily has emphasized on the production of food grains since the government policies promote not only exports but also food security and sustenance of rural and vulnerable sections of the economy. In recent times, India has witnessed a sharp increase in the productivity of food grains, but the underlying factors are of grave concern since issues such as suboptimal production, underutilization of resources and inability to adopt advanced technologies remain unacknowledged. The present study delves into various aspects of the production of food grains across 30 Indian states and emphasizes upon measuring the efficiency of food grain production across the 30 states on the basis of the non-parametric technique of Data Envelopment Analysis (DEA). The evaluation also considers economies of scale. The results highlight the fact that the Indian states are about 21% inefficient in terms of food grain production with the average efficiency score being 0.79. The methodology adopted for this study incorporates crucial factors such as usage of land area, usage of fertilizers and allocation of bank credit to carry the analysis forward. The present study has also aimed at providing certain policy recommendations for the policymakers in this regard so that

Growth, Poverty and Developmental Aspects of Agriculture, 11–25

Published under exclusive licence by Emerald Publishing Limited
doi:10.1108/978-1-83608-076-320241002

the states can sustainably improve their efficiency in terms of the production of food grains.

Keywords: Food grains production; Data Envelopment Analysis (DEA); efficiency; fertilizers; bank credit; Indian states

Introduction

Indian agriculture, its production and different dimensions have long been areas of discussion among international policymakers. Although an agrarian economy, with respect to its contribution to the national employment and share of gross domestic product, India found its production insufficient to feed its own population since independence and had to import food grains. The holistic change in the agricultural scenario since the late 1960s through Green Revolution changed the scenario to a certain extent. The decades of the 1970s and 1980s witnessed a lot of changes in the agricultural landscape, which made India, as a whole, self-sufficient in food grain production; however, not all of the Indian states fit under that category. From 1947 to the 1990s, Indian agriculture went through a lot of changes, which can be referred to as 'Revolutions': starting from Green Revolution, to land reform, to increasing irrigational facilities, to helping out the farmers by proving Minimum Support Prices (MSPs), to the Government itself purchasing food grains from the farmers and ensuring proper redistribution of food grains across the nation by means of Public Distribution System. All of these measures contributed to an increase in the efficiency of the production of food grains and development of Indian agriculture under the government control as a part of its policies.[1] In order to become a self-reliant economy, India had to ensure at least 4% annual growth in agriculture so that it could reduce its Balance of Payments deficit and become an exporter of food grains.

Post-liberalization period, that is, the decades of 1990s, witnessed major shifts in all sectors of India, and agriculture was no exception. The modernization effects of liberalization, privatization and globalization traced through all the economic spheres which caused not only major shift in the patterns of agricultural production but changes in the efficiency as well. From an importer of food grains in the 1960s, India became an exporter of food grains in the decade of 1990s. However, over the decade, the importance of agriculture has been replaced by the service sector – especially, from the perspective of contribution to national income. In addition, the other contributions of agriculture such as being the largest sector of employment generation, food grains supply for feeding the world's most populous nation and earning foreign currency for the nation by way of exports cannot be ignored. General Agreement on Tariffs and Trade (GATT) and World Trade Organization (WTO) agreements and their effects of Indian agriculture along with urbanization of the Indian economy at a very fast pace

[1]Various forms of subsidies, including loan-waive and interest free loans, were given to this sector for this purpose.

have brought modernization in Indian agriculture, and the patterns of food grain production have shifted from food crops to that of cash crops. However, the aspect of producing at the efficient level cannot be ignored; rather, it should be prioritized for the sake of balanced development of the nation. Even the food crop production should not get the second rank, especially the experience of COVID gave that lesson, and a huge proportion of population are still dependent upon free or subsidized food crops across the nation for survival. All these discussion actually highlight the importance of agriculture in Indian scenario along with an emphasis on its smooth and efficient production for the sake of the economy.

The state agricultural plans by National Bank for Agriculture and Rural Development (NABARD) divided states into several agro-climatic zones and then took necessary measures for each of the zones – a much more needed as well as comprehensive measure for the holistic development of the Indian agriculture (Chatterjee, 2017a). Now, the balanced as well as fully potential development of agriculture do not limit itself to the development of food crops only. It consists of commercial crops, export-oriented crops, fisheries, pisciculture and horticulture as well. However, the basic focus on food grain production and increasing the production by efficiently utilizing the existing facilities for the sake of self-feeding and export earning has remained unchanged.[2] In fact, it has become more important whether India has to become a five-trillion economy – where agriculture, especially food grain production, is going to act as the base and work as the backbone of the economy for a balanced cum regional development. Indian economy has been enjoying a good run of economic growth and recovering fast from the shock of COVID, and the dividend of favorable working-age population can only be reaped, multiplied and spread across different sectors if agriculture is properly developed and strengthened, and for achieving that, at the highest possible efficient level, the production of food grain gets even more importance and should be accordingly emphasized through analysis and research.

Following the recent demographic shifts and urbanization process, the increasing demand for food grains has necessitated for a large-scale production of food grains across the Indian states. Emerging market and developing economy (EMDE) like India has dominated the world trade for a decade since India is a key exporter of agricultural products, and India has emerged as a major agricultural exporter of food grains such as rice, oilseeds, castor oil, coffee, cashew, fresh vegetables and sugar in the international market.[3] It is noteworthy to mention that India achieved a record high food grain production in 2021–2022, estimated at 315.7 million tonnes. Not only in terms of exports, but this large-scale production of food grains also caters for a huge percentage of the

[2]Self-feeding stands for feeding the people of Indian economy.

[3]The Ministry of Agriculture and Farmers Welfare has revealed that the share of exports by agriculture and allied activities has accounted for 11.90% of the total merchandise exports of India in 2021–2022. There has been a significant increase by 20.79% in the level of exports in 2021–2022 as compared to 2020–2021, which was mainly due to a surge in production of food grains in India as put forward by the Department of Agriculture and Farmers Welfare.

vulnerable section of Indian society. The long-lasting nature of poverty has been addressed by introducing several beneficial schemes such as PM Garib Kalyan Anna Yojana through which the states and Union Territories have been allocated about 1,118 LMT food grains during the COVID-19 pandemic to pull out the vulnerable section from a state of extreme penury. On the contrary, it is also worth mentioning that India has not emphasized much upon boosting its primary exportable sector in terms of adopting advanced technologies for agricultural production, and hence, India has not been able to achieve its optimal growth trajectory as the farmers are unable to exploit the existing technologies (Murthy et al., 2009). Even though the volume of food grain production has significantly increased over the years, the growth has not increased as expected. Compared with earlier measures, the government has initiated certain sector-specific programs such as Pradhan Mantri Fasal Bima Yojana (PMFBY), MSP Scheme, Restructured Weather Based Crop Insurance Schemes (RWBCIS) in collaboration with the respective state governments along with setting up Common Services Centre (CSC), Krishi Vigyan Kendra and Farmer Producer Organizations (FPOs) to get the farmers trained and make them utilize advanced technologies on farming purposes across the country.

There exists a wide gap in terms of food grain production across Indian states. It has been observed that in 2020–2021, Uttar Pradesh accounted for the highest food grain production, whereas Puducherry reported the least production among the Indian states.[4] India, with its vast agricultural landscape and diverse palette of food grains, plays a significant role in securing global food security. Agriculture in India has also shaped the enhanced rural economy that we observe in recent times, providing sustenance to the vulnerable or relatively poorer section of India under the act of National Food Security (Chatterjee, 2019a; Gulati et al., 2023). With such advancements, it has become necessary for the states to adopt technological innovations (Chatterjee, 2022, 2023; Thakur et al., 2022) just to optimize resource utilization. With evolving dynamics of cultivation, the stakeholders are also getting concerned with the issue of sustainability of Indian agriculture and allied activities, leading to the adoption of efficient methods for production (Chatterjee, 2019b, 2022; Kumar et al., 2022).

In this regard, various aspects of production should be acknowledged since the productivity of both food grains and non-food grains depends on certain crucial factors. Indian agriculture itself is a vulnerable sector with the workforce constituting the major percentage of India's informal economy as well as it is vulnerable in terms of changes in both endogenous and exogenous factors. Indian states experience varying natural resources and production mechanism. The disparity among the levels of productivity among the Indian states is mainly due to difference in climate, pattern of land use, usage of fertilizers and high yielding variety (HYV) seeds and adoption of modern and efficient modes of cultivation.

---

[4]According to the *Handbook of Statistics on Indian States* 2022–2023, Uttar Pradesh accounted for production of 58,106.9 thousand tonnes of food grains and Puducherry for production of 50.7 thousand tonnes of food grains in 2020–2021. Refer to the report for the detailed data on other states.

The rural agricultural sector heavily depends upon the climatic conditions while cultivating major food grains in India. IMD data reveal increasing instances of extreme weather conditions due to climate change. Apart from that, the Agricultural Census 2020–2021 shows that even though India has about 157 million hectares of net sown area, the land cannot be optimally utilized due to fragmentation of land holdings, which affects the economics of scale for food grains productivity.[5] In addition to that, while initiatives such as Digital Agriculture Mission (DAM) promotes technological adoption and efficiency of production, lack of awareness, affordability and low digital literacy seem to be hindrances to such initiatives (Chatterjee, 2017b).[6] Several other factors should also be acknowledged in this regard, and the present literature draws motivation from these issues.

The present study aims at investigating the efficiency of Indian states in terms of food grain production. It also emphasizes on discussing the degree of utilization of resources in order to achieve a higher or optimal growth trajectory. In this context, this chapter adopts the non-parametric technique of Data Envelopment Analysis (DEA) to find the efficiency of the production of food grains across Indian states and analyze accordingly in terms of their efficiency scores. The efficiency analysis also helps to determine economies of scale and identify the underutilized factor of production which in turn may provide insight to the policymakers. The study looks to find how far Indian states are productive or efficient in the production of food grains along with good-performing or worst-performing states. The study also emphasizes on the determining factors for improving the efficiency of the production of food grain across Indian states along with the aim towards providing policy insight for improving the present scenario by focusing on the driving forces behind agriculture. Such studies, covering all the Indian states as well as applying DEA, are missing in the existing literature – a lacuna which the present study looks to fill up.

Brief Review of the Existing Literature

Improving efficiency across various aspects of Indian agriculture needs to be addressed while exploring the ample existing literature. Various studies and extensive research have been done in this regard highlighting issues such as efficiency in agricultural productivity, optimal resource allocation and utilization, adoption of advanced agricultural tools and technologies and fluctuating nature of credit. In this context, numerous studies have addressed the vitality of efficiency in Indian agriculture, encompassing several dimensions such as technical efficiency (TE), allocative efficiency (AE) and economic efficiency (EE) as well as Stochastic Frontier Analysis (SFA).

[5] Agricultural Census 2020–2021 reveals that it is only less than 1.2 hectares that an average land holding size is.

[6] DAM emphasizes upon adoption of technologies such as artificial intelligence, blockchain, remote sensing and GIS technology and its implementation in agriculture and allied activities.

Mathur and Raju (2018), Kanesh et al. (2021), and Singh and Mehta (2021) have assessed the TE of Indian agriculture as well as of Indian states and revealed for a fact that the agricultural productivity in various regions across India is severely lagging behind, and they have also highlighted significant variations in efficiencies across regions. These studies have considered some of the key factors influencing such disproportional efficiency like access to credit, pattern of land use and adoption of advanced technology. Nassiri and Singh (2008) have adopted the DEA technique to investigate the energy use efficiency for paddy crop in Punjab, India, whereas Chauhan et al. (2006) have attempted at discussing efficiencies of farmers with regard to rice production in the state of West Bengal, India, by taking up DEA approach. A few seminal works by Lau and Yotopoulos (1971) and Mahajan et al. (2014) have pioneered the application of efficiency analysis in the context of exploring Indian agriculture, employing a stochastic frontier model to assess the technical efficiency. Jana et al. (2023) have analyzed the efficiency for 20 Indian states and pointed out that the selected states were only 60% efficient. They have also considered literacy level as a factor influencing efficiency. Byerlee and Hussain (1997) have argued that institutional reforms and improved market access can enhance AE of agriculture. In addition to these, several other notable works have also contributed in this regard. Parikh et al. (1993) have analyzed the impact of price distortions and input subsidies on efficiency across different regions and for various crops. A study by Kumar and Pandey (2014) has examined the impact of adoption of advanced technologies such as drip irrigation upon enhancing water use efficiency and patterns of resource allocation. On the other hand, the study by Sharma and Sharma (2010) has adopted the measures of technical and scale efficiency to analyze the efficiency of small industries in India. The present literature has inquired into the research gaps of the existing literature and developed a study incorporating all the necessary elements lacking in the existing literature. The following sections thoroughly discuss the technical efficiency analysis on the basis of DEA in an output-oriented framework for the Indian states. Chatterjee (2023) has shown how climate change impacts the productivity of agriculture and forestry and thus affects the migration. Chatterjee (2022) has shown how people value the forestry in dry lands and Sundarbans and thus shows how in the presence of over forest-dependency, agriculture serves as the second source of livelihood. How the lack of ground water coupled with dryness of the weather jointly contribute to the scarcity in the agricultural goods. Chatterjee (2019a) has vividly shown why and how aridity of the soil and the area have affected agriculture in the dry land areas of West Bengal, and thus, forestry became the dominant source of livelihood. Chatterjee and Chatterjee (2021) have shown how far people are willing to contribute from their own pockets for the development of agriculture in the dry lands of West Bengal. This was the first use of contingent valuation method (CVM) for the sustainable development of agriculture for the dry land areas of West Bengal. This study found that people were really willing to contribute from their pockets for the development of agriculture, and the eagerness for contribution was immensely influenced by rainfall, level of education, availability of irrigation, facility of multiple cropping and so on. Despite the abject poverty in the area, peoples' willingness for the development of agriculture was really significant and influencing for the policy-setting bodies. Previous studies on the backwardness or underdevelopment of agriculture in the backward areas further motivated us for this study.

The main motivation of the study generates from the fact that no study has considered all the Indian states in a single study by incorporating the major factors that do influence the agricultural performance. Again, the application of DEA is almost absent for judging the agricultural performance in Indian circumstances, especially by combining all the states. This study becomes even more relevant for using the latest available data – another aspect that gives the study an edge since it covers the dual effects of both COVID and post-COVID effects in Indian agriculture and then goes for comparing the interstate comparison of efficiency – which is missing in the recent literature. The COVID period has witnessed several subsidized facilities for the Indian agriculture – including higher support price, purchase of food grains by the governments, interest less loans and waiver or delay of previous loans – are only few. How far those facilities have been influential in increasing the efficiency of the states are by and large reflected here in this study – all these aspects are immensely helpful not only to the state agricultural bodies but also to the national authorities as well. The absence of such studies has, by and large, worked as the prime motivation behind this study.

Data and Methodology

Data

For the present study, state-level data have been obtained from *Handbook of Statistics on Indian States* 2022–2023 (www.rbi.org.in), published by the Reserve Bank of India. The concerned study has adopted the non-parametric DEA method to measure and analyze the relative efficiencies of Indian states in terms of food grain production. The input–output model has incorporated various key factors such as total production of food grains (1,000 tonnes), cropping intensity (in per cent), area utilized for total food grains (1,000 hectares), per hectare consumption of fertilizers (kg. per hectare) and bank credit to agriculture by scheduled commercial banks (Rs. crore) to analyze the varying efficiency in food grain production across the states.[7]

[7]The data reflect the fact that there is a significant difference among the states in terms of utilizing the factors necessary for food grain production. The data for 2020–2021 show that West Bengal had the highest cropping intensity of 193.2%, whereas Manipur had the least at 100%. In addition to that, Uttar Pradesh had a significant area of 19,547 thousand ha. for food grain production, whereas Puducherry had just about 20.3 thousand ha. In terms of consumption of fertilizer, Tamil Nadu tops the chart with a consumption of 312 kg. per ha., and several states have reported no usage of fertilizers. The data reveal that Tamil Nadu had received a whooping amount of Rs. 252,472 crores as credit by scheduled commercial banks, whereas Arunachal Pradesh had received only Rs. 224 crores as bank credit. This reflects a disparity in terms of utilization of inputs.

Farell (1957) has provided with an extensive study on TE and AE. Following from that, TE can be interpreted in terms of a difference between optimal level of potential output and the observed level of output. The criterion for this study is to maximize the level of output, that is, the food grain production, given the level of resources or inputs in an output-oriented framework. The model also follows from the benchmark model of DEA (Charnes et al., 1978) under constant returns to scale. The following model has considered the respective states as decision-making units (DMU) and has considered of factors such as total food grain production, cropping intensity, area for total food grains, consumption of fertilizers and bank credit to measure and assess the efficiency levels of the states in terms of the production of food grains. Following the non-parametric output-oriented DEA framework using the linear programming (LP), the present model aims at maximizing the observed output to its optimal level by considering effective alterations of the input levels specified. Hence, the model pursues an output-oriented Constant Returns to Scale (CRS) structure to determine the efficiencies. The present study has considered of the datasets for the year 2020–2021 to assess the technical efficiency of the Indian states.

Model of Technical Efficiency

The concerned model for this study has considered N states or DMUs ($N = 30$). It is assumed that each of the N states produces 'm' outputs utilizing 'n' inputs. Furthermore, we have considered each state i produces output bundle $y^i = (y_{1i}, y_{2i}, \ldots, y_{mi})$ by employing input bundle $x^i = (x_{1i}, x_{2i}, \ldots, x_{ni})$. In addition, if we assume the technology characterized by CRS condition, the production possibility set can be expressed as:

$$T^{CRS} = \left\{ (x, y) : x \geq \sum_{j=1}^{N} \mu_j x^i;\ y \leq \sum_{j=1}^{N} \mu_j y^i;\ \mu_j \geq 0;\ (j = 1, 2, \ldots, N) \right\}$$

We can now define the output-oriented TE measure for firm i producing output y^i employing the input bundle x^i under CRS technology following the LP problem. Let ϕ^* be the optimum value of ϕ such that $(x^i, \phi y^i) \in T^{CRS}$ and $y^* = \phi^* y^i$. The LP problem can be specified as:[8,9]

$$\text{Max } \phi$$

$$\text{Subject to} \sum_{j=1}^{N} \mu_j y_{tj} \geq \phi y_{ti};\ (t = 1, 2, \ldots, m)$$

$$\sum_{j=1}^{N} \mu_j x_{kj} \leq x_{ki};\ (k = 1, 2, \ldots, n)$$

$$\mu_j \geq 0;\ (j = 1, 2, \ldots, N)$$

---

[8] ϕ is considered as the output-oriented measure of TE of any firm i. That is the degree of change in y to reach the threshold y^*. The objective is to maximize ϕ.

[9] Following from the assumptions for production technology, we can specify $x \geq \sum \mu_j x^i$ and $y \leq \sum \mu_j y^i$.

The output-oriented TE of firm i can be defined as:

$$TE_O^i = TE_O(x^i, y^i) = \frac{y^i}{\phi^* y^i} = \frac{1}{\phi^*}$$

The present study has modeled a 1-output and 4-inputs framework under production technology with CRS.[10] The model has considered the production of food grains as the output and cropping intensity, area, consumption of fertilizer and bank credit as inputs for production.[11] The estimated efficiency scores mentioned in the following section lies between 0 and 1 with efficiency score 1 representing optimal utilization of resources in terms of maximizing food grain production, whereas an efficiency score closer to 0 highlights underutilization of resources by the states and suboptimal level of food grain production.

Results and Analysis

This section particularly highlights the estimated measure of TEs for food grain production for 30 states in India. The primary findings of this study have been presented in Panel I of Table 2.1, which basically provides the descriptive statistics of the variables taken into consideration for measuring TE. The results suggest that on an average, the Indian states produce 10,352.11 thousand tonnes of food grains. It is also evident from the results that on an average, the cropping intensity for the states is 148.7% and the average land area utilized for food grain production is 4,336.62 thousand hectares. Additionally, the states on an average consume 110.51 kg. per hectare of fertilizers along with Rs. 55,617.3 crores bank credit by scheduled commercial banks on an average for agricultural activities. The results also reveal for a fact that there exists a significant disparity among the states regarding the allocation of the concerned factors that affect the productivity of food grains, which can be observed from the maximum and minimum values of the variables in consideration.

The key findings of the present study, that is, the respective measures of the technical efficiencies for the states, have been highlighted in Panel II of Table 2.1. The depicted results suggest that under the output-oriented measure of TE as well as under the CRS assumption, the states on an average are at a 0.793 level of efficiency in terms of the production of food grains. It suggests that the states are about 21% inefficient in terms of optimizing their output levels for food grains, which could have been reduced if the given inputs were utilized efficiently. The output-oriented measure of technical efficiency shows that states like Chhattisgarh, Madhya Pradesh, Meghalaya, Nagaland, Punjab, Rajasthan and Uttar Pradesh have the highest efficiency score of 1.000 that reflects the fact that these states have achieved the optimal level of the production of food grains. It has been

---

[10]We know from input-oriented TE measure, $TE_I^i = \frac{\theta^* x^i}{x^i} = \theta^*$ and under the CRS assumption $\theta^* = \frac{1}{\phi^*}$, that is, $TE_I^i = TE_O^i$.

[11]Specifications of the concerned variables have been mentioned at the beginning of Section III.

Table 2.1. Descriptive Statistics of the Factors and Efficiency Scores of Indian States.

Panel I: Descriptive Statistics of the Factors Affecting the Efficiency of Production

Variable	Mean	Std. Dev.	Min	Max
Food grain production	10,352.11	12,996.77	50.7	58,106.9
Cropping intensity	148.7	25.73	100	193.2
Area of food grain production	4,336.62	5,168.67	20.3	19,547
Bank credit	55,617.3	66,017.79	224	252,472
Fertilizers	110.51	82.33	0	312

Panel II: Efficiency Scores of Indian States in Terms of Production of Food Grains

States	Efficiency Score	States	Efficiency Score
Andhra Pradesh	0.617	Manipur	0.972
Arunachal Pradesh	0.829	Meghalaya	1
Assam	0.865	Mizoram	0.787
Bihar	0.760	Nagaland	1
Chhattisgarh	1	Odisha	0.775
Goa	0.571	Puducherry	0.547
Gujarat	0.461	Punjab	1
Haryana	0.949	Rajasthan	1
Himachal Pradesh	0.473	Sikkim	0.692
Jammu and Kashmir	0.398	Tamil Nadu	0.817
Jharkhand	0.845	Telangana	0.689
Karnataka	0.636	Tripura	0.792
Kerala	0.705	Uttar Pradesh	1
Madhya Pradesh	1	Uttarakhand	0.756
Maharashtra	0.544	West Bengal	0.944
Average efficiency	0.793		

Source: Authors' estimation on the basis of state-level data from *Handbook of Statistics on Indian States* 2022–2023 (www.rbi.org.in).

observed from the recent trends that stable economic returns, free power for irrigation, use of HYV seeds and fertilizers and procurement of heavy bank credit have led Punjab to be an exceptional performer in terms of agricultural production. Out of all food grains, paddy cultivation has been given utmost importance in Punjab for which 3.13 million ha. out of 3.59 million ha. of land under cultivation has been utilized. Similarly, the output of food grain in Uttar Pradesh has reached 45 million tonnes in recent times. A multipronged strategy has been adopted by the state government to boost the yield of the food grain production. Giving distinct priorities to Ravi and Kharif crops, the state has been able to improve farmers' income by allowing the sector to receive subsidies under government schemes on fertilizers and to import advanced machineries from abroad along with direct small-scale credit from the rural banks. Madhya Pradesh has excelled in the production of grams and pulses. The year 2020–2021 has observed a 97.5% growth in production of urad. Under Pradhan Mantri Kisan Samman Nidhi (PMKISAN) and Mukhya Mantri Kisan Kalyan Yojana, marginal farmers have been allocated with a financial assistance of Rs. 6000 and Rs. 4000, respectively. The promotion of agricultural mechanization under National Mission on Agriculture Extension (ATMA) has resulted in a hike in the yield of food grain production in Madhya Pradesh. The results from this study also reveal that some of the north-eastern states have been performing exceptionally well in terms of production. The scheme of National Mission for Sustainable Agriculture by the Government of India has enhanced the efficiency in terms of usage of water and technology as well as in the mentioned north-eastern states. On the contrary, the results suggest that states like Jammu and Kashmir (0.398), Gujarat (0.461), Maharashtra (0.544) and Puducherry (0.547) are some of the major states that critically lag behind in terms of efficiency in the production of food grains and could optimize the use of resources to increase the level of production. In the regions of Jammu and Kashmir, high cropping intensity is heavily constrained by the lack of advanced sowing technology resulting in low growth of food grain productivity. On the other hand, Gujarat faces disruptions in terms of the pattern of land use since the distribution of land over there is much fragmented and lack of technology in the face of groundwater contamination leads to the lower yield of food grains due to serious salinity of the soil. Apart from that, Haryana (0.949) and West Bengal (0.944) have also shown satisfactory performance in terms of the production of food grains as these two states are considered as major food grain producer in the economy.

Moreover, Fig. 2.1 represents the state-wise technical efficiency measures for food grain production and evidently shows that there is a noteworthy difference among the states regarding the efficiency measure, which also reflects the fact that there has been underutilization of the existing resources by the states as a result of which there has been suboptimal production.

Following from the model of output oriented TE, the results have exhibited CRS as well as highlighted the state-wise efficiency measures or efficiency scores for the production of food grains. The results have revealed that it is due to improper use of specified inputs that the states are only 79% efficient in terms of production. The remaining 21% can be accounted for if the states that are lagging behind can efficiently adopt the technology and utilize the resources to improve their present level of production. It is to be noted that the role of credit is evolving and with the foreign participation the states can introduce broader ways for the farmers and the investors as well.

Fig. 2.1. Efficiency Score of Indian States in Food Grains Production. *Source:* Authors' estimations on the basis of state-level data from *Handbook of Statistics on Indian States* 2022–2023.

Conclusion

The vast literature on efficiency measures in terms of agricultural production has put forward several dimensions which have shown how varying the factors of production might affect the productivity of agricultural output as well as to what extent the underlying factors are responsible for the low productivity of the agricultural sector. The present study has explored the aspect of efficiency in productivity of food grains across 30 Indian states. The study has considered of several crucial factors, such as cropping intensity, area utilized under food grain production, consumption of fertilizers and bank credit, and also analyzed their impact on the production of food grains across the states in India. The present chapter has adopted the output-oriented model of technical efficiency to derive and explain the respective efficiency scores of the Indian states.

The study has thoroughly examined the efficiency of all the Indian states in terms of the production of food grains, and the results have revealed that there exists a significant level of inefficiency among most of the Indian states. Only a handful of states like Chhattisgarh, Madhya Pradesh, Meghalaya, Nagaland, Punjab, Rajasthan, and Uttar Pradesh have exhibited the highest efficiency score of 1 under CRS, and along with that, Haryana (0.949) and West Bengal (0.944) have also been major contributors of food grain production efficiently. Apart from that, the remaining states have been producing at an inefficient or suboptimal level. Additionally, the descriptive statistics has shown the extreme inequality among the states in terms of utilization of the inputs while optimizing the level of the production of food grains.

In an economy that suffers from severe current account deficit, where malnutrition among people is a common phenomenon and food subsidy along with supply of free food grains are the basic necessities and demands –inefficient food grain production from the states is a worrying indication. India is one of the prime exporters of food grains across the developed economies; hence, inefficiency in the production from several states does raise questions on part of the proper policy-setting and demands a revision, reframe or even implementation of new agricultural policies with proper infrastructural and institutional facilities that would ensure efficiency, on part of the states, in the production of food grains. India has been doing its best to become a five-trillion economy over the past decade and thereby become an global economic powerhouse, and neglecting food grain production would not only hamper the achievement of such goals but also would put question marks before the basic issues of sustainable development. Hence, proper government interventions with holistic policy-changes, keeping in purview the question of region-specific needs, are the needs of the agricultural scenario at present in India.

References

Byerlee, D., & Hussain, S. S. (1997). Assessing the impact of technical change on labor use: A critical overview of models and methods. *Agricultural Economics*, *16*(3), 157–173.

Charnes, A., Cooper, W. W., & Rhodes, E. (1978). Measuring the efficiency of decision making units. *European Journal of Operational Research*, *2*(6), 429–444.

Chatterjee, N. (2017a, March). Willingness to pay for drinking water in some selected dryland areas of West Bengal: A contingent valuation approach. *Vidyasagar University Journal of Economics*, *19*.

Chatterjee, N. (2017b). An assessment of infrastructural facilities in the dryland areas of West Bengal. In R. C. Das (Ed.), *Social, health and environmental infrastructures for economic growth*. IGI Global.

Chatterjee, N. (2019a, March). Valuation of forestry in selected dryland areas of West Bengal: A contingent valuation approach. *Economic Affairs*, *64*(1), 173–183.

Chatterjee, N. (2019b, December). Use of renewable resources and sustainable livelihood in Indian Sunderbans: An attempt through contingent valuation method. *Vidyasagar University Journal of Economics*, *22*.

Chatterjee, N. (2022, December). Households' willingness to pay for improved water supply system in the dryland areas of West Bengal – An estimation using double-bounded dichotomous-choice model. *Economic Affairs*, *67*(05), 787–795.

Chatterjee, N. (2023). Climate change and labour force migration in Indian Sundarbans: An econometric exploration. *Inclusive*, *2*(24). December 2023 – January 2024.

Chatterjee, N., & Chatterjee, T. (2021, January). Agricultural practices and sustainable livelihood in the drylands of West Bengal: An assessment through contingent valuation method. In A. Bhat & J. Kachroo (Eds.), *Agribusiness development planning and management* (pp. 169–185). New Delhi Publishers. https://doi.org/10.30954/NDP.agribusiness.2020.14

Chauhan, N., Mohapatra, P., & Pandey, K. (2006). Improving energy productivity in paddy production through benchmarking – An application of data envelopment analysis. *Energy Conversion and Management*, *47*, 1063–1085.

Farell, M. J. (1957). The measurement of productive efficiency. *Journal of the Royal Statistical Society*, *120*(3), 253–290.

Gulati, A., Paroda, R., Puri, S., Narain, D., & Ghanwat, A. (2023). Food system in India. Challenges, performance and promise. In J. von Braun, K. Afsana, L. O. Fresco, & M. H. A. Hassan (Eds.), *Science and innovations for food systems transformation*. Springer.

Jana, S., Nayek, D., Roy, P., & Jana, S. (2023). State-wise efficiency of agricultural production in India: Data envelopment analysis approach. *Scholars Journal of Economics, Business and Management*, *10*, 130–135.

Kanesh, S., Wilson, C., Khanal, U., & Santhirakumar, S. (2021). How productive are rice farmers in Sri Lanka? The impact of resource accessibility, seed sources and varietal diversification. *Heliyon*, *7*, e07398. https://doi.org/10.1016/j.heliyon.2021.e07398

Kumar, N., Chhokar, R. S., Meena, R. P., Kharub, A. S., Gill, S. C., Tripathi, S. C., Gupta, O. P., Mangrauthia, S. K., Sundaram, R. M., Sawant, C. P., Gupta, A., Naorem, A., Kumar, M., & Singh G. P. (2022). Challenges and opportunities in productivity and sustainability of rice cultivation system: A critical review in Indian perspective. *Cereal Research Communications*, *50*, 573–601.

Kumar, A., & Pandey, R. (2014). Impact of drip irrigation on water use efficiency: A review. *International Journal of Engineering Science Invention*, *3*(7), 40–46.

Lau, L., & Yotopoulos, P. (1971). A test of relative efficiency and application to Indian agriculture. *The American Economic Review*, *61*, 94–109.

Mahajan, V., Nauriyal, D. K., & Singh, S. P. (2014). Technical efficiency analysis of the Indian drug and pharmaceutical industry: A non-parametric approach. *Benchmarking: An International Journal*, *21*(5), 734–755.

Mathur, R., & Raju, S. (2018). Efficiency in food grains production in India using DEA and SFA. *The Central European Review of Economics and Management*, *2*, 79. https://doi.org/10.29015/cerem.589

Murthy, D. S., Sudha, M., Hegde, M. R., & Dakshinamoorthy, V. J. A. E. R. R. (2009). Technical efficiency and its determinants in tomato production in Karnataka, India: Data envelopment analysis (DEA) approach. *Agricultural Economics Research Review*, *30*(2), 259–270.

Nassiri, S., & Singh, S. (2008). Study on energy use efficiency for paddy crop using data envelopment analysis (DEA) technique. *Applied Energy*, *86*, 1320–1325.

Parikh, A., Ali, F., & Bailey, D. (1993). Price distortions and resource use efficiency in a Pakistani province. *European Review of Agricultural Economics*, *20*(1993), 35–47.

Sharma, S., & Sharma, M. (2010). Analyzing the technical and scale efficiency of small industries in India: State-wise cluster study. *Measuring Business Excellence*, *14*(2), 54–65.

Singh, R. K., & Mehta, P. K. (2021). Technical efficiency and its determinants in Indian agriculture: An analysis using data envelopment analysis. *Indian Journal of Agricultural Economics*, *76*(2), 253–269.

Thakur, A. K., Singh, R., Gehlot, A., Kaviti, A., Aseer, R., Suraparaju, S. K., Natarajan, S. K., & Sikarwar, V. S. (2022). Advancements in solar technologies for sustainable development of agricultural sector in India: A comprehensive review on challenges and opportunities. *Environmental Science & Pollution Research*, *29*, 43607–43634.

Chapter 3

Fostering Agricultural Growth Sustainability: The Critical Role of Bank Financing in Nigeria

Olalekan Charles Okunlola[a] *and Olumide Abiodun Ayetigbo*[b]

[a]University of Derby, UK
[b]Department of Business Administration, National Open University of Nigeria, Nigeria

Abstract

This study examines the impact of deposit money banks' loan on agricultural sustainability in Nigeria for the period 1985–2017. The study adopted the ARDL bonding test to determine the short- and long-run effect of deposit money banks' loan on agricultural sustainability in Nigeria. It found that bank loans have a negative impact on agricultural sustainability in both the short and long run. Similarly, deposit money banks' interest rate also has an inverse relationship on agricultural growth in Nigeria in both the long and short run. The study concludes that bank finance is very impactful in stimulating agricultural growth within the context of the Nigerian economy. Therefore, the study recommends that the Nigerian government should increase agricultural financing and lower interest rates in order to achieve food sufficiency and promote agricultural sustainability.

Keywords: Bank financing; bank interest rate; ACGSF scheme; agricultural sustainability; ARDL; Nigeria

Introduction

Agriculture in Nigeria was the main source of foreign exchange earnings even before the discovery of crude oil in commercial quantities in Nigeria. The main national priority in Nigeria is to ensure food security for its vast population, which is estimated to be around 200 million (FMARD, 2016). Agriculture is a

Growth, Poverty and Developmental Aspects of Agriculture, 27–40

Published under exclusive licence by Emerald Publishing Limited
doi:10.1108/978-1-83608-076-320241003

significant sector of the Nigerian economy, contributing about 20% to the country's gross domestic product (GDP) and employing more than 60% of the labor force. It has also been observed that more than half of the population in developing countries is involved in traditional agriculture and subsistence farming (World Bank, 2017). However, the sector faces a range of challenges, including inadequate financing, poor infrastructure, and limited access to modern technologies and markets. In terms of financing, agriculture in Nigeria is largely underfunded, with only a small fraction of commercial banks' loan portfolio allocated to the sector. This lack of financing leads to limit the farmers' ability to invest in their businesses, purchasing modern equipment and technologies, and accessing markets beyond their immediate communities. Nigeria despite her increased population over the years has not been able to sufficiently produce enough quantum of food items to feed the growing population; this is because of poor funding, the subsistence nature of agricultural practices, and lack of mechanized equipment such as tractors, plow, and harvester that could help boost the productivity in the sector (Enyim et al., 2013; Obilor, 2013).

To address this issue, the Nigerian government has implemented several initiatives aimed at increasing financing for the agriculture sector. For instance, the Central Bank of Nigeria has set up various loan schemes, including the Anchor Borrowers' Program (ABP) and the Commercial Agriculture Credit Scheme (CACS), to provide farmers with affordable credit facilities. In addition, there are private sector initiatives such as agricultural financing companies and impact investors that are providing financing and technical support to small and medium-sized farmers. One of such organizations is the Nigeria Incentive-Based Risk Sharing System for Agricultural Lending (NIRSAL), which was set up to de-risk agricultural lending and encourage commercial banks to lend to the sector.

However, despite these efforts, financing remains a significant challenge for many smallholder farmers in Nigeria, particularly those in rural areas. This is because of the high-interest rates on loans, lack of collateral and inadequate knowledge about financial management. While there have been efforts to increase financing for agriculture in Nigeria, more needs to be done to support smallholder farmers, who make up most agricultural producers in the country. This includes the provision of affordable credit facilities, improved infrastructure, access to modern technologies and markets, and capacity building to enhance financial literacy and management.

Deposit money banks through its credit policy act as an engine that promotes growth in various sectors of the economy by channeling resources to the real sector (Akpansung & Gidigbi, 2014). It has also been established that sufficient financing of agricultural projects will not only promote food security but also enhance the entrepreneurship performance of our young investors, which is expected to impact positively on the investible sectors of the economy through improved agricultural production of goods and services (Imoisi et al., 2012; Nnamocha & Eke, 2015; Udih, 2014).

In the literature, there has been an argument regarding the relationship between bank's financing and agricultural growth (Akinlo & Okunlola, 2022; Apanisile & Okunlola, 2014; Imoisi et al., 2012; Ngong et al., 2022; Okunlola & Akinlo, 2021; Olowofeso et al., 2017). Some studies found a positive and significant relationship between bank loans and agricultural development (Akinlo &

Okunlola, 2022; Apanisile & Okunlola, 2014; Imoisi et al., 2012; Ngong et al., 2022; Nnamocha & Eke, 2015; Okunlola & Akinlo, 2021; Olowofeso et al., 2015). However, some other studies have observed no relationship between bank credit and agricultural output (Okafor et al., 2016; Olowofeso et al., 2017). Obilor (2013) is of the opinion that money banks favor credit and advances to other sectors other than the agricultural sector; as a result, banks charge farmers high-interest rate knowing fully well that farmers will not be able to meet up. The argument continues in the literature, and there is no consensus as to the relationship between bank financing and agricultural growth sustainability. Okunlola and Akinlo (2021), Apanisile and Okunlola (2014), and Akinlo and Okunlola (2022) all found that Agricultural Credit Guarantee Scheme Fund (ACGSF) and Commercial Bank Credit to Private Sector have positive effects on growth and poverty reduction. These studies, however, studied a single component of bank's financing and couldn't give detailed information on its effect on agricultural growth sustainability and poverty reduction.

Many studies have empirically examined the impact of bank financing on agricultural growth sustainability (Akinriola & Okunola, 2017; Akudugu, 2016; Anh et al., 2020; Chandio et al., 2017; Florence & Nathan, 2020; Kumar et al., 2017; Ngong et al., 2020; Osabohien, Adeleye, et al., 2020; Osabohien et al., 2018; Osabohien, Mordi, et al., 2020; Rehman et al., 2017). Some studies have reported a positive relationship only in the long run (Chandio et al., 2017; Florence & Nathan, 2020; Ngong et al., 2020; Nnamocha & Eke, 2015), while others reported no functional relationship between deposit money banks' loan and agricultural growth sustainability (Olowofeso et al., 2017; Rehman et al., 2017). Other studies also found a positive relationship in both long and short run between bank financing and agricultural development (Akinlo & Okunlola, 2022; Anh et al., 2020; Osabohien, Mordi, et al., 2020). For instance, Akudugu (2016) noted that the relationships between formal credit, informal credit, and farm size on one hand and agricultural productivity on the other are nonlinear in nature. The interactions of formal credit with informal credit, informal credit with farm size, and formal and informal credit with farm size have significant relationships with agricultural productivity. Similarly, Nnamocha and Eke (2015) found a long-run relationship between bank credit and agricultural output.

There is no consensus in the literature regarding the effect of bank financing on agricultural growth sustainability in the long and short run. This is why it is instructive to investigate the long and short-run effect of bank financing on agricultural growth sustainability. This further helps Nigeria government and policymakers to create policy environment that will help sustain the growth that is been experienced in the agricultural sector of the economy. This study departs from other studies as it focuses on bank financing on agricultural growth sustainability in Nigeria. Bank financing ranges from financing from deposit money banks and public agricultural banks and international financing in the country. This chapter is organized as follows: Section 2 presents the methods and materials, section 3 presents and discusses the findings of the study, and section 5 contains the conclusion.

Materials and Methods

To analyze the impact of bank financing on agricultural growth sustainability, the study relies on the Barro regression (Barro & Sala-i-Martin, 2003). It states that economic growth is a function of financial system development and a vector of other indicators. This will be represented in a functional form as:

$$y = f\{\text{FSD}, \chi^{n}_{i=1}\} \tag{3.1}$$

where y stands for economic growth, FSD is financial sector development, and χ is the vector of other factors affecting economic growth. The study adopts and adapts Eq. (3.1), then it becomes

$$A_y = f\{\text{BF}, \chi^{n}_{i=1}\} \tag{3.2}$$

"y" economic growth is being substituted by agricultural growth A_y, while financial system development (FSD) is being substituted by bank financing. Eq. (3.2) then becomes:

$$\text{AGS}_t = \alpha_0 + \beta_1\text{FIN}_t + \beta_2\text{GFCF}_t + \beta_3\text{AMACH}_t + \beta_4\text{ALN}_t + \beta_5\text{ALAB}_t + \mu_t \tag{3.3}$$

Where AGS is agricultural growth, BF is bank financing measured by deposit money banks' loan to the agricultural sector (BCA), interest rate (INT), and ACGSF to representing domestic financing and foreign direct investment (FDI) representing foreign financing, ALAB is agricultural labor, ALAN is agricultural land area, AMACH is agricultural machinery and capital (proxied with gross fixed capital formation).

This study, to achieve the main goal (i.e., to determine the short- and long-run effects of financing on agricultural growth sustainability in Nigeria) describes the characteristics of the data used. Table 3.1 shows the descriptive statistics of the variables like AGS, BCA, ALAB, ALAN, AMACH, , FDI, and capital (proxied with gross fixed capital formation) used in this study from 1985 to 2017.

The study conducted initial tests of stationarity and descriptive statistics. The results are in order and can be presented by request. This study adopts a cointegration method to demonstrate any long-term relationship between bank financing and agricultural growth sustainability in Nigeria over the period 1985–2017. The study chapter applies the autoregressive distributed lag (ARDL) bounds testing approach developed by Pesaran et al. (2001) to investigate the primary objective of the research work stated in objective 3.

It is expected that the coefficient of the FDI, BCA, and ACGSF be positive. The main reason for this is that FDI and domestic financing are incentives for farmers to produce and export more output. Mainly, they have the potential to reduce the cost of production and enhance efficiency and productivity in the agricultural sector. However, where agriculture operates majorly at the subsistence level as obtained in the Nigerian economy, the response to incentives by the operators in the sector might be prolonged and weak. Consequently, agricultural development schemes may not produce the expected outcomes, particularly in the short run.

Table 3.1. Estimated ARDL Long-Run Coefficients.

Dependent Vrb: AGDP ARDL (2, 1, 0, 0, 1, 0, 0)				
ARDL Bound test for cointegration				
MODEL	*F*-Statistic			
F(BCA, ALAB, ALAN, AMACH, GFCF, FDI)	10.557			
Critical Values	Upper Bound		Lower Bound	
$K=6$; $n=37$				
10%	2.94		1.99	
5%	3.28		2.27	
Estimated ARDL Long-Run Coefficients				
Variable	Coefficient	*t*-statistic	Prob.	
BCA	−0.0013	−4.6224	0.000	
ALAB	−0.0982	−3.4908	0.003	
ALAN	0.0040	0.7519	0.463	
AMACH	0.3371	3.4763	0.003	
GFCF	−0.0220	−0.6862	0.502	
FDI	−0.0046	−0.2511	0.806	
C	11.4066	6.2502	0.000	
Error Correction Representation for the Selected				
Variable	Coefficient	S. E	*t*-Stat	*P*-value
D(BCA)	−0.0008	0.0002	−4.2380	0.001
D(AMACH)	0.0461	0.0511	0.9033	0.381
D(ALAB)	−0.1208	0.0512	−2.3623	0.031
D(ALN)	0.0049	0.0070	0.7068	0.490
D(GFCF)	−0.0271	0.0416	−0.6520	0.524
D(FDI)	−0.0057	0.0226	−0.2507	0.805
CointEq(−1)*	−1.2593	0.1131	−11.1300	0.000
EC = AGDP − (−0.0013*BCA −0.0982*ALAB + 0.0040*ALN + 0.3371*AMACH − 0.0220*GFCF − 0.0046*FDI + F10)				
R-squared	0.8098	Mean dependent var	0.0603	
Adjusted *R*-squared	0.7838	S.D. dependent var	0.0795	
S.E. of regression	0.0369	Akaike info criterion	−3.6179	
Sum squared resid	0.0300	Schwarz criterion	−3.4243	
Log-likelihood	51.0327	Hannan-Quinn criter.	−3.5622	
Durbin Watson stat	2.4027			

Source: Authors' computation, 2023.

Note: Here Optimum lag is selected based on Schwarz Bayesian Criterion. CointEq(−1)* – significant results.

Based on the bounds-testing approach proposed by Pesaran and Shin (1999) and Pesaran et al. (2001), adopted by Okunlola and Ayetigbo (2024), Okunlola et al. (2023), Okunlola (2019), Okunlola and Akinlo (2021), the long-run relationship is given by the equation:

$$\begin{aligned}\Delta AGDVP_t = {} & \beta_0 + \textstyle\sum_{i=0}^{q_1}\gamma_i\Delta AGDVP_{t-1} + \sum_{i=0}^{q_1}\partial_i\Delta BF_{t-1} + \sum_{i=0}^{q_2}\delta_i\Delta ALAB_{t-1} \\ & + \textstyle\sum_{i=0}^{q_3}\rho_i ALAN + \sum_{i=0}^{q_4}\vartheta_i\Delta AMACH_{t-1} + \sum_{i=0}^{q_5}\theta_i\Delta GFCF_{t-1} \\ & + \textstyle\sum_{i=0}^{q_6}\omega_i\Delta FDI_{t-1} + \varphi_0 AGDVP_{t-1} + \varphi_1 BF_{t-1} + \varphi_2 ALAB_{t-1} \\ & + \varphi_3 ALAN_{t-1} + \varphi_4 AMACH_{t-1} + \varphi_5 GFCF_{t-1} + \varphi_6 FDI + \epsilon_t\end{aligned} \tag{3.4}$$

where q is the optimal lag length and Δ refers to the first difference of variables. The hypothesis for testing the existence of any long-run co-integration among the variables is given thus:

$$H0.\ \varphi 1 = \varphi 2 = \varphi 3 = \varphi 4 = \varphi 5 = \varphi 6 = \varphi 7 = 0$$
$$H1.\ \varphi 1 \neq 0,\ \varphi 2 \neq 0,\ \varphi 3 \neq 0,\ \varphi 4 \neq 0,\ \varphi 5 \neq 0,\ \varphi 6 \neq 0,\ \varphi 7 \neq 0$$

This states the joint null hypothesis of no co-integration against the existence of co-integration between agricultural output growth and the set of explanatory variables. Given that there is co-integration, the short-run model is stated as:

$$\begin{aligned} \text{AGDVP}_t = & \beta_0 + \sum_{i=0}^{q_1} \gamma_i \Delta\text{AGDVP}_{t-1} + \sum_{i=0}^{q_1} \partial_i \Delta\text{BF}_{t-1} + \sum_{i=0}^{q_2} \delta_i \Delta\text{ALAB}_{t-1} \\ & + \sum_{i=0}^{q_3} \rho_i \text{ALAN} + \sum_{i=0}^{q_4} \vartheta_i \Delta\text{AMACH}_{t-1} + \sum_{i=0}^{q_5} \theta_i \Delta\text{GFCF}_{t-1} \\ & + \sum_{i=0}^{q_6} \omega_i \Delta\text{FDI}_{t-1} + \lambda\text{ECM}_{t-1} \end{aligned} \quad (3.5)$$

The coefficients γ_i, ∂_i, δ_i, ρ_i, ϑ_i, θ_i, and ω_i denote the short-run dynamics of the variables, while the coefficient $\phi 1(i = 1, 2, 3, 4, 5, 6)$ denote the long-term dynamics. The term λ is the coefficient of correction in disequilibrium.

Results and Discussion

Long- and Short-Run Effect of Commercial Bank's Loan to Agricultural Sector on Agricultural Growth

The presence of long-run relationships between the variables is tested by using Eq. (3.4). Table 3.1 reports the results of the ADRL bounds co-integration tests. It shows the calculated F-statistics when the regression is normalized on the agricultural output. The Wald tests (F tests) for the joint null hypothesis that the coefficients of the lagged variables in the level form are zero (no co-integration between the variables), and the results of the calculated F-statistics and the values for both upper and lower bound are presented in Table 3.1. The computed F-statistic (10.557) is higher than the upper critical bound at 5% (3.28) and 10% (2.94) critical values as indicated in Table 3.1. This provided evidence to reject the null hypothesis of no co-integration at 5% and 10% significance level for the growth model. The study, therefore, concludes from the ARDL bounds test that there is a long-run relationship among the variables. The study obtains the long-run and short-run dynamic parameters for the variables once a long-run co-integration relationship has been established. The empirical results of the long-run model are presented in Table 3.2.

The study estimates Eq. (3.4) using the following ARDL (2, 1, 0, 0, 1, 0, 0) specification. This estimated the short and long-run relationship between commercial bank loans to the agricultural sector and agricultural growth in Nigeria. The result in Table 3.1 shows that deposit money banks' loan to the agricultural sector which is a measure of bank finance hurts agricultural growth and sustainability in the long run. This indicates that a unit increase in commercial banks' loan to the agricultural sector will reduce agricultural growth and sustainability

Table 3.2. Estimated ARDL Long-Run Coefficients.

Dependent Vrb: AGDP ARDL (1, 1, 1, 2, 2, 2, 1)		
ARDL Bound test for cointegration		
MODEL		*F*-Statistic
F(INT, ALAB, ALAN, AMACH, GFCF, FDI		9.827
Critical Values	Upper Bound	Lower Bound
	K=6; *n*=37	
10%	2.94	1.99
5%	3.28	2.27

Estimated ARDL Long-Run Coefficients			
Variable	Coefficient	*t*-statistic	Prob.
INT	−0.015	−4.642	0.001
ALAB	−0.030	−1.983	0.079
ALN	−0.003	−0.583	0.574
AMACH	0.583	10.056	0.000
GFCF	−0.110	−2.969	0.016
FDI	0.031	2.168	0.058
C	7.201	7.458	0.000

Error Correction Representation for the Selected ARDL Model ARDL(1, 1, 1, 2, 2, 2, 1)a				
Variable	Coefficient	S. E	*t*-Stat	*P*-value
D(INT)	−0.007	0.001	−5.669	0.000
D(ALAB)	−0.285	0.026	−10.82	0.000
D(ALN)	0.009	0.003	2.713	0.024
D(AMACH)	−0.007	0.037	−0.2	0.846
D(GFCF)	−0.224	0.03	−7.585	0.000
D(FDI)	0.075	0.01	7.529	0.000
CointEq(−1)*	−1.257	0.106	−11.822	0.000

EC = AGDP −(−0.0145*INT −0.0030*ALAB −0.0030*ALN + 0.5832*AMACH −0.1099*GFCF + 0.0313*FDI + 7.2012)

R-squared	0.9548	Mean dependent var	0.0602
Adjusted *R*-squared	0.9294	S.D. dependent var	0.0795
S.E. of regression	0.0211	Akaike info criterion	−4.5937
Sum squared resid	0.0071	Schwarz criterion	−4.1099
Log-likelihood	69.7186	Hannan-Quinn criter.	−4.4544
Durbin-Watson stat	2.4815		

Source: Authors' computation, 2021.

Note: a. Selected based on Schwarz Bayesian Criterion. CointEq(−1)* = significant results.

by −0.0013 and vice versa. This result is, however, contrary to the a priori expectations as it shows a negative impact of commercial banks' credit to the agricultural sector on agricultural growth. Some scholarly works have demonstrated the positive effects of deposit money banks' credit to the agricultural sector on agricultural growth (Dong et al., 2010; Duy, 2012). While many studies established a direct and significant impact of commercial banks' credit to the agricultural sector on agricultural growth (Bashir et al., 2010; Ekwere & Edem, 2014; Saleem & Jan, 2011), many studies have argued that an indirect relationship exists between them (Ahmad, 2011; Raza & Siddiqui, 2014). The finding of this study aligns with Obilor (2013) who revealed that commercial bank credit to the agriculture sector and agricultural produce prices hurt agricultural productivity and growth sustainability in Nigeria.

The short-run result in Table 3.1 shows that a bank's credit to agriculture has a negative impact, as the same as what is obtained in that long run. This meant that in both the long and short run, the bank's credit to the agricultural sector has a negative and significant effect on agricultural GDP. This finding to a large extent supports the work and findings of Obilor (2013). Although this negative impact is contrary to a priori expectations, some studies in the literature have found a similar result. This negative impact may be due to vices like corruption which will not ensure that these bank credits are received by those farmers that need the funds. Nigeria is a very complex country, given the fact that the nation has weak institutions; the bank credit may not be properly channeled to where it would be more productive in the agricultural sector. Also, in the short-run result, the coefficient of agricultural labor has a similar sign to the long-run result. This indicates that in both the long- and short-run, agricultural labor negatively impacts agricultural output.

Long- and Short-Run Effects of Commercial Banks' Interest on Agricultural Growth Sustainability

This study also investigated the impact of deposit money banks' interest rate on agricultural development in Nigeria; the presence of long-run relationships between the variables was tested. The computed *F*-statistic (9.827) is higher than the upper critical bound at 5% (3.28) and 10% (2.94) critical values as indicated in Table 3.2. This provided evidence to reject the null hypothesis of no co-integration at 5% and 10% significance level for the growth model. The study, therefore, concludes from the ARDL bounds test that there is a long-run relationship among the variables.

The study estimates Eq. (3.3) using the following ARDL (1, 1, 1, 2, 2, 2, 1) specification. The result in Table 3.2 shows that deposit money banks' interest rate, which is also a measure of bank finance, hurts agricultural output in the long run. This indicates that a unit increase in commercial banks' interest rate will reduce agricultural output by 0.015 and vice versa. This result is however in line with the a priori expectations and other scholarly works, as it shows a negative impact of deposit money banks' interest rate on agricultural output. Some scholarly works have demonstrated the negative effects of deposit money banks' interest rate on agricultural GDP (Mashinini et al., 2019). One channel through which interest rate impacts agricultural output is agricultural investment (Asekome & Ikojie, 2018). Asekome and Ikojie (2018) established that deposit interest rate has a positive impact on agricultural investment, growth, and development.

On the other hand, the South African department of agriculture, forestry, and fisheries (2010) found that low-interest rates have not curbed farm debt but rather increased them. They found that this was due to the increasing prices of farming equipment and input even with low agricultural product prices which forced farmers to seek more loans to get the required capital for production.

Similarly, in the short run, the coefficient of interest rate is negative. This indicates that in the short run, interest rates negatively impact agricultural output in Nigeria. Specifically, a unit increase in interest rate will reduce agricultural output by 0.007 in the short run (see Table 3.2). This impact, though negative and significant, value is smaller than the value in the long run. In this model, FDI has a positive and significant effect on agricultural output in the long and short run.

Furthermore, the estimated coefficient of the lagged error correction term (ECM) (−1) in Table 3.2 has the expected negative sign, and it is also statistically significant at 0.01 level. This implies that deviations from the long-term growth rate in agricultural output adjust quickly. The speed of adjustment is high; it shows that about 125.7% of the disequilibrium within the model in the previous year is subsequently corrected in the current year. In summary, the findings in this subsection validate the premise that the extent of bank financing, as measured by the interest rate, exerts a detrimental impact on the agricultural Gross Domestic Product (GDP) of Nigeria. To be more specific, a rise in the interest rate is associated with a contraction in agricultural GDP both in the short and long run. Therefore, it becomes imperative to establish a robust and efficacious monetary policy in order to foster and sustain growth within Nigeria's agricultural sector. This policy should strive to maintain an equilibrium interest rate to ensure the enduring development of the agricultural sector and, consequently, bolster Nigeria's overall economic well-being.

Long- and Short-Run Effects of ACGSF on Agricultural Growth Sustainability

The computed *F*-statistic (14.09671) is higher than the upper critical bound at 5% (3.28) and 10% (2.94) critical values as indicated in Table 3.3. This provided evidence to reject the null hypothesis of no co-integration at 5% and 10% significance level for the model. The study, therefore, concludes from the ARDL bounds test that there is a long-run relationship among the variables. Then the study obtains the long-run and short-run dynamic parameters for the variables since the long-run co-integration relationship has been established.

The study estimates Eq. (3.3) using the following ARDL (3, 1, 2, 2, 2, 2, 2) specification. The estimated coefficients of the long-run relationship of the impact of ACGSF on agricultural output in Nigeria. The results in Table 3.3 show that the ACGSF, which is also a measure of bank finance, has a positive effect on agricultural output in the long run. This indicates that a unit increase in the ACGSF will reduce agricultural output by 0.095 and vice versa. This result is consistent with the a priori expectations and other scholarly works, as it shows a positive impact of the agricultural credit guarantee scheme funded on agricultural output. Some scholarly works have demonstrated the negative effects of the ACGSF on agricultural GDP (Obilor, 2013; Dangusau, 2022). Obilor (2013) revealed that ACGSF produced a significant positive effect on agricultural productivity in Nigeria.

Table 3.3. ARDL Bound Test for Co-integration.

Dependent Vrb: AGDP ARDL (3, 1, 2, 2, 2, 2, 2)a				
MODEL			*F*-Statistic	
F(ACGSF, ALAB, ALAN, AMACH, GFCF, FDI)			14.09671	
Critical Values	Upper Bound		Lower Bound	
	K=6; *n*=37			
10%	2.94		1.99	
5%	3.28		2.27	
Estimated ARDL Long-Run Coefficients				
Variable	Coefficient	*t*-statistic	Prob.	
ACGSF	0.0945	3.6392	0.022	
ALAB	–0.0648	–3.426	0.027	
ALN	0.0168	3.4830	0.025	
AMACH	0.0451	0.3524	0.742	
GFCF	0.0343	4.5957	0.010	
FDI	0.0644	–0.2511	0.805	
C	7.8020	6.5408	0.003	
Error Correction Representation				
Variable	Coefficient	S. E	*t*-Stat	*P*-value
D(ACGSF)	0.3093	0.1328	2.3286	0.027
D(ALAB)	–0.0917	0.0405	–2.2662	0.031
D(ALN)	0.7659	0.2899	1.6416	0.011
D(AMACH)	0.4078	0.1486	2.7446	0.011
D(GFCF)	0.2472	0.0920	2.6883	0.497
D(FDI)	0.3788	0.1431	2.6466	0.013
CointEq(–1)*	–0.4780	0.1973	–2.4232	0.022
EC = AGDP – (0.3093*ACGSF – 0.0917*ALAB + 0.7659*ALN + 0.3371*AMACH – 0.0046*FDI)				
R-squared	0.809759	Mean dependent var	0.06028	
Adjusted *R*-squared	0.783817	S.D. dependent var	0.07947	
S.E. of regression	0.036949	Akaike info criterion	–3.6179	
Sum squared resid	0.030036	Schwarz criterion	–3.4243	
Log-likelihood	51.03267	Hannan-Quinn criter.	–3.5622	
Durbin-Watson stat	2.402742			

Source: Authors' computation, 2023.

Note: a. Selected based on Schwarz Bayesian Criterion. CointEq(−1)* – significant results.

Also, Sulaimon (2021) evaluated the threshold analysis of ACGSF on agricultural GDP in Nigeria found a positive effect. He suggested that a sustained increase in the value of agricultural loans guaranteed and the inclusiveness of more smallholder farmers who dominate the Nigerian agricultural space will translate into a robust contribution of the scheme to agricultural performance in Nigeria. Similarly, Awotide et al. (2015), Osabohien, Mordi, et al. (2020), and Osabohien, Adeleye, et al. (2020) give credence to the significant impact of credit financing on agricultural productivity in the country.

In the short run, however, the result in Table 3.3 shows a positive but not significant impact of ACGSF on agricultural output. Unlike Osabohien, Adeleye, et al. (2020), who found a positive and significant impact of agricultural financing on agricultural performance in both the long and short run, this study found a positive but insignificant result ACGSF on agricultural GDP in the short run. The result in Table 3.3 indicates that the impact of ACGSF on agricultural output in the short run in Nigeria is insignificant. Many studies have looked at the effect of ACGSF on agricultural output in Nigeria (Obilor, 2013; Dangusau, 2022). The ACGSF was introduced by the federal government in 1977 to address the challenge of agricultural financing deficit and achieve food security in Nigeria and to encourage formal financial institutions to increase and sustain lending to agriculture (Eyo et al., 2020).

The results regarding the effect of FDI on agricultural growth sustainability vary depending on the model. For example, in the model that examined the effect of ACGSF on agricultural growth (Table 3.3), FDI only has a positive and significant impact in the short term. On the other hand, in the model that explored the relationship between interest rates and agricultural growth, FDI has a positive and significant effect. Based on these findings, it is not possible to confidently conclude that international financing is beneficial for the sustainability of agricultural growth in Nigeria. The agricultural sector in Nigeria is not particularly attractive to foreign investors, as the level of FDI in this sector remains relatively low compared to other sectors of the Nigerian economy, despite an increase in recent years. According to the Nigerian Investment Promotion Commission (NIPC), the agricultural sector received $4.4 billion in FDI between 2013 and 2017. In 2018, the sector attracted $200 million in FDI, which accounted for 3% of the total FDI inflow into Nigeria. The primary sources of FDI in Nigeria's agricultural sector are China, the United States, the United Kingdom, the Netherlands, and India. These investments primarily focus on agro-processing, commercial farming, and livestock production.

Conclusion

The study has examined whether bank financing had significantly improved agricultural growth sustainability in Nigeria. The study employs ARDL bound testing approach to test the short- and long-run dynamics between bank financing and agricultural growth sustainability in Nigeria.

The study found that indicators for bank financing have significant effect on agricultural growth sustainability in Nigeria. For instance, increased deposit money banks' credit to the agricultural sector hurts agricultural growth sustainability in both the long and short run. Although this finding is contrary to a priori knowledge, the high level of corruption and institutional weaknesses have been considered as a possible cause of this result. Also, the study found that the interest rate hurts growth sustainability in Nigeria. In other words, a higher interest rate will reduce agricultural GDP in both the long and short run in Nigeria. And lastly, the study also establishes that the ACGSF positively impacts agricultural growth sustainability only in the long run in Nigeria.

The study concludes that bank financing is very impactful in stimulating agricultural growth sustainability in Nigeria. This is shown in the impact of deposit money banks' credit to the agricultural sector, deposit money banks' interest rate, and ACGSF on agricultural growth sustainability in Nigeria.

References

Ahmad, N. (2011). Impact of institutional credit on agricultural output: A case study of Pakistan. *Theoretical and Applied Economics*, *10*(563), 99–120.

Akinlo, A. E., & Okunlola, O. C. (2022). On the interaction between export promotion and agricultural growth in poverty reduction in Nigeria: Empirical evidence for the period 1980–2016. *Journal of Economic Development*, *47*(1), 109–134.

Akinriola, O. O., & Okunola, A. M. (2017). Performance analysis of Nigerian agricultural credit guarantee scheme: Bounds test approach to cointegration. *Journal of Development and Agricultural Economics*, *12*(2), 75–83.

Akpansung, A. O., & Gidigbi, M. O. (2014). Recent banking reforms in Nigeria: Implications on sectoral credit allocation and economic growth. *International Journal of Business and Social Science*, *5*(13), 91–104.

Akudugu, M. A. (2016). Agricultural productivity, credit, and farm nexus in Africa: A case study of Ghana. *Agricultural Finance Review*, *76*(2), 288–308.

Anh, N. T., Gan, C., & Anh, D. L. T. (2020). Does credit boost agricultural performance? Evidence from Vietnam. *International Journal of Social Economics*, *47*(9), 1203–1221.

Apanisile, O. T., & Okunlola, O. C. (2014). Growth effect of export promotion and nonoil output in sub-Saharan Africa (1970–2014). *Emerging Economy Studies*, *3*(2), 1–17.

Asekome, M. O., & Ikojie, E. S. (2018). An empirical investigation on impact of interest rates on agricultural investment in Nigeria. *Journal of Economics and Business*, *1*(1), 71–82.

Awotide, B. A., Abdoulaye, T., Alene, A., & Manyong, V. M. (Eds.). (2015). *Impact of access to credit on agricultural productivity: Evidence from smallholder cassava farmers in Nigeria*. https://doi.org/10.22004/ag.econ.210969

Barro, R. J., & Sala-i-Martin, X. (2003). *Economic growth*. The MIT Press.

Bashir, M. K., Mehmood, Y., & Hassan, S. (2010). Impact of agricultural credit on productivity of wheat crop: Evidence from Lahore, Punjab, Pakistan. *Pakistan Journal of Agricultural Sciences*, *47*(4), 405–409.

Chandio, A. A., Jiang, Y., Gessesse, A., & Dunya, R. (2017). The nexus of agricultural credit, farm size and technical efficiency in Sindh, Pakistan: A stochastic production frontier approach. *Journal of the Saudi Society of Agricultural Sciences*, *18*(3), 348–353.

Dangusau, M. S. (2022). Agricultural credit guarantee scheme fund in historical context: Evidence from Nigeria. In E. F. Wamboye & B. Fayissa (Eds.), *The Palgrave handbook of Africa's economic sectors* (pp. 135–154). Springer International Publishing. https://doi.org/10.1007/978-3-030-75556-0_6

Dong, F., Lu, J., & Featherstone, A. M. (2010). *Effects of credit constraints on productivity and rural household income in China*. Center for Agricultural and Rural Development, Iowa State University.

Duy, V. Q. (2012). *The role of access to credit in rice production efficiency of rural households in the Mekong Delta, Vietnam*. Center for Asian Studies Discussion Paper No. 284. Centre for International Management and Development Antwerp.

Ekwere, G. E., & Edem, I. D. (2014). Evaluation of agricultural credit facility in agricultural production and rural development. *Global Journal of Human-Social Science Research*, *14*(3), 19–26.

Enyim, O. B., Ewno, E. N., & Okoro, O. T. (2013). Banking sector credit and the performance of the agricultural sector in Nigeria. *European Journal of Scientific Research*, *23*(2), 35–55.

Eyo, E. O., Nwaogu, M. A., & Agenson, M. E. (2020). Agricultural credit guarantee in Nigeria and the uncertainties of the macroeconomic environment. *International Journal of Economics and Financial Issues*, *10*(20), 20–29.

Federal Ministry of Agriculture and Rural Development (FMARD). (2016). *The agriculture promotion policy (2016–2020): Building on the successes of the ATA, closing key gaps, policy and strategy document*. Abuja.

Florence, N., & Nathan, S. (2020). The effect of commercial bank's credit on agricultural growth in Uganda. *African Journal of Economic Review*, *3*(1), 162–175.

Imoisi, A. A., Sogules, I. W., & Ekpeyoung, B. I. (2012). An appraisal of credit facilities on agricultural output and productivity in Nigeria: 1970–2010. *British Journal of Humanities and Social Sciences*, *7*(2), 24–32.

Kumar, A., Mishra, A. K., Saroj, S., & Joshi, P. (2017). Institutional versus non-institutional credit to agricultural households in India: Evidence on impact from a national farmers' survey. *Economic Systems*, *41*(3), 420–432.

Mashinini, S. M., Dlamini, G. S., & Dlamini, V. D. (2019). The effects of monetary policy on agricultural output in Eswatini. *International Journal of Economics and Financial Research*, *55*, 94–99. https://doi.org/10.32861/ijefr.55.94.99

Ngong, C. A., Onyejiaku, C., Fonchamnyo, D. C., & Onwumere, J. U. J. (2022). Has bank credit really impacted agricultural productivity in the central African economic and monetary community? *Asian Journal of Economics and Banking*. https://doi.org/10.1108/AJEB-12-2021-0133

Ngong, C. A., Thaddeus, K. J., & Onwumere, J. U. (2020). Banking sector development and agricultural productivity in central African economic and monetary community (CEMAC). *Journal of Agricultural Science and Technology*, *10*, 68–82.

Nnamocha, P. N., & Eke, C. N. (2015). Bank credit and agricultural output in Nigeria (1970–2013): An error correction model (ECM) approach. *British Journal of Economics, Management & Trade*, *10*(2), 1–12.

Obilor, S. I. (2013). The impact of commercial banks' credit to agriculture on agricultural development in Nigeria: An econometric analysis. *International Journal of Business, Humanities and Technology*, *3*(1), 85–95.

Okafor, I. G., Ezeaku, H. C., & Ugwuegbe, U. S. (2016). Relationship between deposit money bank credit and economic growth in Nigeria under a VAR G-causality environment. *IOSR Journal of Economics and Finance*, *7*(2), 41–46.

Okunlola, O. C. (2019). Political regime types and economic development in Nigeria: Significance of conflict and corruption. *Journal of Interdisciplinary Economics*, *31*(2), 183–216. https://doi.org/10.1177/0260107918769539

Okunlola, O. C., & Akinlo, E. A. (2021). The impact of export promotion schemes on agricultural growth in Nigeria. *African Journal of Economic Review*, *9*(1), 60–68. Article 1.

Okunlola, O. C., & Ayetigbo, O. A. (2024). Impact of agricultural financing on agricultural growth sustainability in Nigeria. *The Journal of Developing Areas*, *58*(3), 171–203.

Okunlola, O. C., Sani, I. U., & Ayetigbo, O. A. (2023). Socio-economic governance and economic growth in Nigeria. *Journal of Business and Socio-Economic Development*. (ahead-of-print) https://doi.org/10.1108/JBSED-03-2023-0019

Olowofeso, O. E., Adeboye, A. A., Adejo, V. T., Bassey, K. J., & Abraham, O. (2017). Agricultural sector credit and output relationship in Nigeria: Evidence from nonlinear ARDL. *CBN Journal of Applied Statistics*, *8*(1), 101–122.

Olowofeso, E. O., Adeleke, A. O., & Udoji, A. O. (2015). Impact of private sector credit on economic growth in Nigeria. *The CBN Journal of Applied Statistics*, *6*(2), 81–101.

Osabohien, R., Adeleye, N., & Alwis, T. D. (2020). Agro-financing and food production in Nigeria. *Heliyon*, *6*, 1–8.

Osabohien, R., Afolabi, A., & Godwin, A. (2018). An econometric analysis of food security and agricultural credit facilities in Nigeria. *The Open Agriculture Journal*, *12*, 227–239.

Osabohien, R., Mordi, A., & Ogundipe, A. (2020). Access to credit and agricultural sector and agricultural sector performance in Nigeria. *African Journal of Science, Technology, Innovation, and Development*, *14*(1), 247–255.

Pesaran, M. H., & Shin, Y. (1999). An autoregressive distributed-lag modelling approach to cointegration analysis. In S. Strøm (Ed.), *Econometrics and economic theory in the 20th century: The Ragnar Frisch Centennial Symposium* (pp. 371–413). Cambridge University Press. https://doi.org/10.1017/CCOL521633230.011

Pesaran, M. H., Shin, Y., & Smith, R. J. (2001). Bounds testing approaches to the analysis of level relationships. *Journal of Applied Econometrics*, *16*(3), 289–326. https://doi.org/10.1002/jae.616

Raza, J., & Siddiqui, W. (2014). Determinants of agricultural output in Pakistan: A Johansen co-integration approach. *Academic Research International*, *5*(4), 30–46.

Rehman, A., Chandio, A. A., Hussain, I., & Jingdong, L. (2017). Is credit the devil in the agriculture? The role of credit in Pakistan's agricultural sector. *The Journal of Finance and Data Science*, *3*, 38–44.

Saleem, M. A., & Jan, F. A. (2011). The impact of agricultural credit on agricultural productivity in Dera Ismail Khan (District) Khyber Pakhtonkhawa Pakistan. *European Journal of Business and Management*, *3*(2), 38–44.

Sulaimon, M. (2021). Agricultural credit guarantee scheme fund (ACGSF) and agricultural performance in Nigeria: A threshold regression analysis. *SSRN Electronic Journal*. https://doi.org/10.2139/ssrn.3973046

Udih, M. (2014). Bank credits and agricultural development: Does it promote entrepreneurship performance? *International Journal of Business and Social Science*, *5*(11), 102–107.

World Bank, World Development Report. (2017). Agriculture for development United States of America. *Monthly Report*, *2*(1), 7–9.

Chapter 4

Do FDI Inflow, Forest Coverage and CO2 Emission Stimulate Agriculture Sector Productivity? Empirical Evidence From ICU Economies

Prasenjit Makur

Calcutta Girls' College, The University of Burdwan, India

Abstract

Foreign direct investment (FDI) has a critical role in boosting agricultural productivity and the growth of emerging economies. The relationship between FDI inflows and environmental factors has not received much attention in identifying its impact on agricultural output. Using annual time series data from 1990 to 2023, this study examines the causal association and short- and long-run effects of FDI inflows, forest coverage and CO2 emissions on the agricultural productivity of the India, China and US (ICU) economies. The autoregressive distributed lag (ARDL) results confirmed that FDI inflows have a significant and positive impact on Indian and Chinese agriculture productivity, whereas CO2 emissions adversely affect US agriculture productivity in the long run. In the short run, CO2 emissions led to agricultural productivity in both China and the US economies. The bound test and error correction model (ECM) result also confirmed the long-run connection and convergence of the equilibrium path among the studied variables except India. The findings of the Granger causality test showed a unidirectional causal link between agriculture productivity and FDI inflows and forest coverage in India and a bidirectional causal link between CO2 emission and agricultural yield and forest coverage and CO2 emission in the Chinese agriculture sector. The study also revealed a unidirectional causal association between forest coverage and agricultural output and between FDI, CO2 emissions and forest coverage in the US agriculture sector. Policymakers were advised to encourage FDI in the agriculture sector and expand the use of environment-friendly technology to decrease carbon

Growth, Poverty and Developmental Aspects of Agriculture, 41–62

doi:10.1108/978-1-83608-076-320241004

emissions and promote forest coverage for sustainable growth and higher agricultural production.

Keywords: FDI Inflows; agriculture productivity; carbon emission; ARDL; forest coverage; ICU

Introduction

The agriculture sector is fundamental to the national economy in both the developed and developing countries. Investing in the agriculture sector is the best strategy to prevent food catastrophes (Timmer, 2010). Linking the agriculture industry with other sectors can help a country grow more quickly, according to a World Development Report (2008). It has been demonstrated that foreign direct investment (FDI) is crucial for fostering economic expansion, improving a nation's technological standing and generating new jobs in emerging nations. Additionally, it has been demonstrated that FDI helps developing nations become more integrated into the world economy by raising the money available for investment. This boosts economic growth, which is necessary to lower poverty and improve living standards. An important industry that depends both directly and indirectly on labour is agriculture. Agricultural practices can harm the environment, but they are also essential for satisfying society's needs for food production and supply. In this regard, the study seeks answers to the following questions: (1) Does FDI inflow contribute to agriculture productivity? (2) Are agriculture and CO2 emissions among the main drivers of environmental degradation in high-agriculture value-added countries? The study attempts to provide answers to these questions to alleviate environmental issues in the top three agriculture value-added nations India, China and the United States (ICU).

The aim of sustainable development and achieving net-zero emissions are hot topics everywhere in the world. According to the Intergovernmental Panel on Climate Change (IPCC) Report 2021, rapid economic growth hurts the climate system, affecting people, systems and entire industries. Currently, in 2022, there are 7.95 billion people in the world; by 2030, that number is predicted to rise to 8.55 billion, as per the United Nations WPP Report 2022. Sustainable development is hampered by the population's ongoing rapid expansion. Meeting the standards outlined in the 17 Sustainable Development Goals (SDGs) is essential to solving environmental issues. The sustainable use of natural resources and the prevention of environmental issues including deforestation, water pollution, desertification, air pollution and global warming are addressed in this context in the 12th, 13th, 14th and 15th SDGs as outlined by the United Nations in Transforming our World: the 2030 agenda for sustainable development. IPCC Working Group III, Climate Change 2014, suggests that to achieve sustainable development and equity, it is imperative to mitigate the effects of climate change and global warming.

Production and consumption in developing nations are growing quickly and becoming more diverse as a result of globalization. Societies are shifting to renewable energy sources like solar and wind, which do not pollute the air and are abundant in nature, because fossil fuels used in the manufacture of commodities are a source of environmental degradation and pose a health risk to humans. Carbon emissions may rise as a result of increased energy use, agricultural output and investments in the energy and agricultural sectors. If anthropogenic greenhouse gas (GHG) emissions continue to cause global warming, the achievement of SDG 2 – the eradication of hunger – could be in jeopardy by 2030. As per statistics from the Food and Agriculture Organization of the United Nations Rome (2022), in 2022, China was the world's largest emitter of GHGs, accounting for 29% of all emissions. At 11% of global emissions, the United States ranked as the second-biggest emitter. Of the 53.8 billion metric tonnes of carbon dioxide equivalent (GtCO2) released in 2022, the six biggest GHG polluters collectively accounted for over 60% of the emissions. As per the Timperley (2019) Carbon Brief Profile report, India ranks third globally in terms of GHG emissions, following the United States and China. Examining India's commitment to lowering its emissions to improve its positive environmental performance is crucial in light of the Paris Agreement, the country's pledge as a stakeholder in the global climate talks and how quickly India ratified the agreement on 2 October 2016, just 1 year later.

Gokmenoglu et al. (2019) advocated that through the provision of food, natural resources, competition and jobs, agriculture can boost economies. Zafeiriou and Azam (2017) suggested that, particularly for developing nations, agricultural pursuits like hunting, forestry and fishing are essential to the generation of economic growth. On the other hand, Pellegrini et al. (2016) pointed out that agriculture may result in increased carbon, land and water footprints. FAO, The State of Food and Agriculture (2019), showed that livestock and agriculture irrigation account for 70% of all water withdrawals worldwide. Additionally, during the main production phase which involves raising cattle, cultivating soil and using agricultural inputs – GHG emissions and carbon footprints rise. With time, the amount of land needed for food production also grows. Ancient agricultural systems and inputs derived from fossil fuels have the potential to harm the environment and obstruct the attainment of the SDGs. As per the report, The Contribution of Agriculture to Greenhouse Gas Emissions, FAO, UN, about 20% of CO2 emissions worldwide come from agricultural output, so lowering agricultural emissions is essential for sustainable development, as evidenced by Gokmenoglu and Taspinar (2018) and Ridzuan et al. (2020), since this tendency will make environmental issues worse worldwide. Following energy (35%) and industry (21%), agriculture is the third-largest contributor to global warming as per GHG emissions from agriculture, forestry and other land use (FAO, UN). According to the study of Ramachandra et al. (2015) and Udemba (2020), the use of fossil fuels in agricultural processes including paddy cultivation, crop and soil burning, land reclamation and other related activities can all directly result in higher emissions.

Food production, leisure, nutrient cycling and the mitigation of climate change are just a few of the environmental and social benefits that come with forest ecosystem services to human society (Costanza et al., 1997, 2017; Millennium Ecosystem Assessment (MEA), 2001; TEEB, 2010). Despite the recognition of these responsibilities, deforestation continues to be a significant problem, especially in tropical regions and African nations (FAO, 2016; Kim et al., 2015; Sloan & Sayer, 2015). According to Costanza et al. (2014), this has led to significant economic losses because it has weakened the direct and indirect benefits provided by the forests. The impact on market pricing is one method used to evaluate the economic worth of direct ecosystem services, also known as direct use value. On the other hand, little research has been done on the internalization of externalities or the indirect economic value of ecosystem services, often known as the indirect use value. Klein et al. (2003) highlighted agricultural pollination, identifying the elements impacting pollination and discovering that as the distance between the forest and the coffee plantation increased, the number of pollinators (social bees) declined. They discovered that these pollinators can lower production failure rates and boost yields. In their native environments, these pollinators support agricultural output. According to Klein et al. (2007), using natural pollinators enhances yields for more than 75% of the major food crops grown worldwide. In a similar vein, Carvalheiro et al. (2010) discovered that, although ants were unaffected, the number of flying pollinators is correlated with the distance to a natural habitat. In their study, Ricketts et al.(2004) investigated the effects of forest-based pollinators on coffee bean yields and quality in Costa Rica.

The distribution of forest ecosystem services has a diversity of effects on agricultural output, such as crop pollination (Carvalheiro et al., 2010; Klein et al., 2003, 2007; Ricketts et al., 2004), soil conservation (Mäder et al., 2002) and nutrient retention (Raudsepp-Hearne et al., 2010). According to Mäder et al. (2002), more biodiversity led to a decreased reliance on chemical inputs, and organic agriculture increased soil fertility. Raudsepp-Hearne et al. (2010) identified 20 ecosystem services related to soil nutrient retention based on spatial land-use patterns. They discovered that, in Canada, the rate of forest cover is positively correlated with soil fertility in terms of phosphorus retention. For instance, in developing countries, forest revenue typically makes up between 22% and 60% of all household income (Angelsen et al., 2014; L'Roe & Naughton-Treves, 2014; Vedeld et al., 2007). Policies aimed at conserving forests usually consider the cost and benefit of forests' capacity to store carbon. Conservation strategies, however, do not include the additional functions of ecosystem services as monetary values. Thus, the monetary worth of ecosystem services has not been considered in forest conservation programmes.

Literature Review

The purpose of this study is to empirically investigate how the agricultural sector of ICU economies is impacted by FDI inflows, CO2 emissions and forestation. Growth has always been a hot topic in economics. The study of Clark (1923), Solow (1956), and Romer (1986), related to various theories, including Keynesian theory, classical theory, neo-classical theory and endogenous growth theory, highlighted the elements influencing growth. Theories have identified the factors that are drawing in more FDI. For example, the Keynesian theory of 1930 placed a strong emphasis on the total amount of money spent on the economy and its output, with economists focusing on government policies and intervention to avert a recession.

Grossman and Krueger (1991) and Shafik and Bandyopadhyay (1992) showed in their groundbreaking research that there is an inverse U-shaped link between various pollution indicators and economic growth. The Environmental Kuznets Curve (EKC) is the definition of the non-linear connection given by Panayotou (1993). The EKC hypothesis's validity is still debatable today. Investigations by Chen et al. (2019) and Prastiyo et al. (2020) claimed that the hypothesis was true, whereas studies by Gill et al. (2018) and Pata and Aydin (2020) claimed the opposite.

Furthermore, the impact of agriculture on environmental degradation has been the subject of current research using a variety of methodologies. Alhassan (2021) advocated that enhanced agricultural output reduces poverty, broadens the distribution of income, ensures food security and fosters economic growth. According to the economic study by Borlaug (2007), agricultural output also has a positive environmental impact. Economic expansion is correlated with agricultural productivity because it increases consumer demand for goods, services and cleaner surroundings. It also strengthens the government's position in enforcing environmental laws. The same reasoning applies to the claim of Burney et al. (2010) that agricultural productivity opposes the expansion of agriculture into forested regions. In addition to lowering pollution and degrading the environment, this promotes forest protection. However, there is rising concern that agricultural output may be a contributing factor to environmental degradation.

According to Selvanathan et al. (2023), in the panel fixed-effect data, emissions of CO2 rise by 0.04% and 0.63%, respectively, for every 1% increase in agricultural productivity and forest cover in Organization for Economic Cooperation and Development (OECD) countries. The findings of Keh et al. (2023) showed that, both in the short and long term, forests have a negative and considerable impact on CO2 emissions. In the long run, however, agricultural land, consumption of energy and FDI all have a favourable impact on CO2 emissions. Yasmeen et al. (2021) showed in their analysis that forestry has a negative influence on CO2 emissions, while agricultural output has a positive impact. Positive direct and moderating impacts of institutional quality on CO2 emissions are observed. According to the causality study, there is a mutually beneficial connection between CO2 emissions and forestry, renewable energy and agriculture. The study by Chidiebere-Mark et al. (2022) revealed a considerable short- and long-run rise in carbon emissions due to net FDI, fertilizer consumption and livestock production. In the meantime, short- and long-term carbon emissions were greatly reduced by the use of renewable energy. The study of Pata (2021) confirmed a long-term connection between renewable energy production, globalization and agricultural practices on Brazil's and China's ecological footprint and carbon dioxide (CO2) emissions. Long-term elasticity showed that while China's environmental strain was greatly reduced by the production of renewable energy, globalization raised pollution indicators. The long- and short-term autoregressive distributed lag (ARDL) estimates of Chandio et al. (2020) confirmed that both agricultural and livestock production have a large beneficial impact on CO_2 emissions in China. On the other hand, power use in agricultural and forest areas harms it, suggesting that both factors lower CO_2 emissions over the long and short terms. The empirical results of Wang et al. (2020) demonstrated that the development of finance, natural resources and economic globalization increases carbon emissions. On the other hand, in the G7 economies, value-added agriculture reduces carbon emissions. The long-run results of Lokonon and Mounirou (2019) suggested the validity of the pollution haven hypothesis for some countries and that of the pollution halo hypothesis for other countries. The findings are also mixed across sub-Saharan African countries for trade openness and urbanization. Following the Toda and Yamamoto methods, Khan et al. (2018) showed that there was a causal relationship between each explanatory variable in Pakistan and the following: forest area, renewable energy, agriculture value added, coal electricity, hydroelectricity and GHG emissions. The relationship between hydroelectricity and GHG emissions, renewable energy and GHG emissions, forest area and GHG emissions, forest area and coal electricity, hydroelectricity and forest area and vegetable area and forest area was found to be unidirectional causative. Sarkodie and Owusu (2016) opined that Ghana's agricultural output increases CO_2 emissions, according to analysis using the ARDL model. On the other hand, Rafiq et al. (2016) found that in 53 nations they chose, agricultural value-added lowers pollution levels using three distinct panel data estimators. Jebli and Youssef (2017) demonstrated that while renewable energy consumption (REC) reduces CO_2 emissions, a 1% increase in agricultural value-added increases CO_2 emissions by 0.36% using the Johansen-Juselius cointegration test and the vector error correction model (VECM) for Tunisia.

Numerous studies have linked the sources of environmental pollution to variables such as non-renewable energy consumption (Liu et al., 2017), trade openness (Balsalobre-Lorente et al., 2019) urbanization (Prastiyo et al., 2020; Ridzuan et al., 2020), FDI (Udemba, 2020), regulatory quality (Olanipekun et al., 2019), economic complexity (Yilanci & Pata, 2020) and globalization (Shahbaz et al., 2017), in addition to the degree of economic development.

Research Gap

From the above overview, there is variation in the research on the relationship between FDI inflows, agriculture productivity and CO_2 emissions. The primary cause of the uneven outcomes is the variations in the legal and economic environments as well as the periods examined in the research. Consequently, policy recommendations cannot be made based solely on the conclusions gained from the current body of research. However, considering shifting economic and regulatory conditions necessitates a thorough analysis of the dynamic link among FDI inflows, agriculture productivity, CO_2 emissions and forestation in ICU economies. Upon reviewing the literature, it is also noteworthy that a significant number of studies lack thorough and insightful analysis, which might potentially guarantee increased stability and robustness in the statistical estimations. To address this problem, the current study uses a variety of rigorous econometric estimations, including the Cumulative Sum (CUSUM) test, in addition to a pertinent set of diagnostic tests to guarantee solid results and a trustworthy conclusion.

Objectives of the Study

In the backdrop of the afore mentioned discussion, the current study has been carried out with the following objectives:

I. To identify the effect of FDI inflows on agriculture productivity in ICU economies.
II. To investigate the influence of CO2 emissions on agriculture productivity in ICU economies.
III. To find the impact of forestation on agriculture productivity in ICU economies.

Research Methodology

Data Source and Variable Description

This analysis contains data for the ICU economies – India, China and the United States – from 1990 to 2023 using an annual secondary data set that was acquired from a variety of sources. Agriculture productivity proxies by gross value-added from agriculture, forestry and fishing (VA_AGR) collected from the FAO's Statistics of the United Nations are single dependent variables that comprise crop cultivation, livestock raising, hunting and fishing in addition to forestry. The net output of a sector after deducting intermediate inputs and adding up all outputs is known as value-added. It is computed without accounting for natural resource depletion and degradation or the depreciation of artificial assets. Carbon emissions (CO2E, metric tonnes), which include emissions from the production of cement and fossil fuels, including carbon dioxide released during gas flaring and the consumption of coal, solid, liquid and gaseous. Coal, fuel oil and natural gas are three independent variables collected from Climate Watch Historical GHG Emissions (1990–2023) in 2023. World Resources Institute, Washington, DC, United States. Forest coverage (FC, sq. km) was collected from the FAO Statistics of the United Nations as an independent variable, which is defined as any territory covered by naturally occurring or artificially planted stands of trees that are at least five metres in length, regardless of whether the trees are productive or not. This definition does not include urban parks and gardens' trees or tree stands inside agricultural production systems, such as fruit plantations and agroforestry systems. FDI inflows (FDI_AGR, US$ million) in the agriculture sector were collected from FAO Statistics of the United Nations, which are the net inflows of capital in that sector targeted to obtain a long-term management stake (10% or more of voting shares) in a business that operates in a different economy than the investor's. In the balance of payments, the sum of equity capital, earnings reinvested, other long-term capital and short-term capital are considered FDI inflows.

Empirical Model

The sample model for the study is shown below, with Y_t as the dependent variable, x_1 to x_3 as the independent variables and ε_t as the error term.

$$Y_t = \beta_0 + \beta_1 x_1 + \beta_2 x_2 + \beta_3 x_3 + \varepsilon_t \tag{4.1}$$

All the variables are converted into natural logarithms (ln) before empirical analysis. Agricultural productivity (lnVA_AGR) is the explained variable of the study. FDI inflows (lnFDI_AGR), CO2 emissions (lnCO2E) and forest cover (lnFC) are the explanatory variables. To investigate the impact of FDI inflows, forest coverage and CO2 emissions on the agriculture sector productivity of the ICU economy, the following equation has been developed:

$$\text{lnVA\_AGR}_{t(I)} = \beta_0 + \beta_1 \,\text{lnFDI\_AGR}_t + \beta_2 \,\text{lnCO2E}_t + \beta_3 \,\text{lnFC}_t + \varepsilon_t \tag{4.2}$$

$$\text{lnVA\_AGR}_{t(c)} = \beta_0 + \beta_1 \,\text{lnFDI\_AGR}_t + \beta_2 \,\text{lnCO2E}_t + \beta_3 \,\text{lnFC}_t + \varepsilon_t \tag{4.3}$$

$$\text{lnVA\_AGR}_{t(\text{US})} = \beta_0 + \beta_1 \,\text{lnFDI\_AGR}_t + \beta_2 \,\text{lnCO2E}_t + \beta_3 \,\text{lnFC}_t + \varepsilon_t \tag{4.4}$$

where lnVA_AGR = Agricultural Value Added for the countries India, China, and the United States, respectively, lnFDI_AGR = Foreign Direct Investment inflows to the agriculture sector, lnCO2E = Carbon Dioxide Emissions, lnFC = Forest Coverage and ε = Error Term.

As a long-term model, we took into consideration Eq. (4.1). We used the bound testing strategy to confirm the long-run relationship in our model. Additionally, the negative and significant value of the error correction term allows us to confirm the long-term relationship. The ARDL model was utilized for analysis; Pesaran et al. (1999) initially employed this technique, and Pesaran et al. (2001) expanded on it again. The ARDL model is the best model to study both the short- and long-run cointegration if the variables are stationary in mix orders, with some variables stationary in $I(0)$ and some stationary in $I(1)$ (Pesaran et al., 2001). This method is being applied at different levels of stationery. Applying the ARDL model is a novel, widely accepted and flexible process. It offers a lot of beneficial circumstances. Incorporating the factors at different levels can lead to a connection (Pesaran et al., 1999).

The ARDL model is superior to the other methods in many ways. First, if the variables are stationary at $I(0)$ and $I(1)$, it doesn't restrict the integration and yields accurate results. Second, we can assess the outcomes of both short- and long-term analyses concurrently, and we can also determine which model is best for small sample data in comparison to other cointegration methods. Third, according to Pesaran et al. (1999), the ARDL model only requires one equation to depict the long-run association. Lastly, Pesaran et al. (2001) clarified that unrestricted error correction model (ECM) is involved in the ARDL model-bound testing technique.

Following earlier research by Dogan and Turkekul (2016), Dogan et al. (2019), Tariq et al. (2019), Seker et al. (2015), and Rehman et al. (2019), the study considers the following equations to determine the long- and short-run relationship among FDI inflows, CO2, forest coverage and agricultural output using the ARDL model:

$$\begin{aligned}\Delta \ln VA\_AGR_t = {} & \alpha_0 + \beta_1(\ln VA\_AGR)_{t-1} + \beta_2(\ln FDI\_AGR)_{t-1} \\ & + \beta_3(\ln CO2E)_{t-1} + \beta_4(\ln FC)_{t-1} + \sum_{t=1}^{p} \varphi_i \Delta \ln VA\_AGR_{t-1} \\ & + \sum_{t=0}^{q} \omega_i \Delta(\ln FDI\_AGR)_{t-1} + \sum_{t=0}^{q} \gamma_i \Delta(\ln CO2E)_{t-1} \\ & + \sum_{t=0}^{q} \delta_i \Delta(\ln FC)_{t-1} + \varepsilon_t \end{aligned} \tag{4.5}$$

where p and q are the lag lengths in the model mentioned above, the variables lnVA_AGR, lnFDI_AGR, lnCO2E and lnFC stand for agricultural productivity, FDI inflows, carbon dioxide emissions and forest coverage, in that order. Additionally, ε_t represents the model's error term, and Δ indicates the difference between the variables. We can proceed to the long-run coefficients as the following models when a long-run cointegration is found from Eq. (4.5).

$$\begin{aligned}\ln VA\_AGR_t = {} & \alpha_0 + \sum_{t=1}^{i=p1} \varphi_i \ln VA\_AGR_{t-1} + \sum_{t=0}^{i=q1} \beta_1(\ln FDI\_AGR)_{t-1} \\ & + \sum_{t=0}^{i=q2} \beta_2(\ln CO2E)_{t-1} + \sum_{t=0}^{i=q3} \beta_3(\ln FC)_{t-1} + \varepsilon_t \end{aligned} \tag{4.6}$$

By constructing the error correction model, the short-run dynamics in the ARDL model can be promptly determined.

$$\begin{aligned}\Delta \ln VA\_AGR_t = {} & \alpha_0 + \sum_{t=1}^{1=p} \varphi_i \ln VA\_AGR_{t-1} + \sum_{t=0}^{i=q1} \omega_i \Delta(\ln FDI\_AGR)_{t-1} \\ & + \sum_{t=0}^{i=q2} \gamma_i \Delta(\ln CO2E)_{t-1} + \sum_{t=0}^{i=q3} \delta_i \Delta(\ln FC)_{t-1} \\ & + \psi\, ECM_{t-1} + \varepsilon_t \end{aligned} \tag{4.7}$$

The error correction term, denoted as ECM_{t-1}, provides the speed of adjustment towards long-run equilibrium. It is best described as follows:

$$\begin{aligned}ECM_t = {} & \ln VA\_AGR_{t-1}\, \alpha_0 - \sum_{t=1}^{i=p1} \beta_1 \ln VA\_AGR_{t-1} - \sum_{t=0}^{i=q1} \beta_2(\ln FDI\_AGR)_{t-1} \\ & - \sum_{t=0}^{i=q2} \beta_3(\ln CO2E)_{t-1} - \sum_{t=0}^{i=q3} \beta_4(\ln FC)_{t-1} \end{aligned} \tag{4.8}$$

The speed of adjustment is indicated by ψ, and all of the coefficients in the short-run equation are coefficients linked to the model's short-run dynamics of convergence to equilibrium.

The Granger causality test is also performed to verify the presence and type of the causal relationship between the variables. Depending on the outcomes of the long-run study, it might be approximated in two different ways. The Granger test (Granger, 1969) can be used to examine the causal link in the short term when there is no cointegration between the variables. However, the standard Granger test is mis-specified when the variables are cointegrated and the error correction method recommended by Engle and Granger (1987) ought to be applied. Given that the variables are cointegrated, the study continues with a Granger causality test using a VECM. For the model's variables, VECM enables the modelling of both the short- and long-term dynamics. While the short-run causality between the variables is examined using the Block Exogeneity Wald test or the VEC Granger causality test, the error correction term of the VECM indicates the direction of long-run causation. Finally, the research performs various diagnostic analyses like normality tests, heteroscedasticity tests, serial correlation tests and CUSUM tests to assess the stability and robustness of the established model.

Results and Discussion

Results of Unit Root Test

The stationary of the data series must be examined in an econometric estimation of time series data. The study used the Augmented Dickey–Fuller test (Banerjee et al., 1993; Dickey & Fuller, 1981) to find out if there were unit roots in the data series. This technique is employed to minimize the possibility of a false relapse. The Dickey–Fuller bound test, which examines the long-term connection in the model while emphasizing stationary variables, indicates that the variables should be stationary at either level $I(0)$ or first difference $I(1)$. None of the variables should be stationary at the second difference $I(2)$ to be estimated under the ARDL model. The measurement of the F statistic yields incorrect findings if the variable is stationary at $I(2)$, which leads to absurdity (Dickey & Fuller, 1981). Therefore, no studied variable should be stationary at $I(2)$ to be estimated under the ARDL model. The variables lnVA_AGR and lnCO2E are found to be stationary at their first differences $I(1)$, whereas lnFDI_AGR and lnFC are revealed to be stationary at level $I(0)$ for all three economies except lnFC of China is stationary at first difference $I(1)$. So, all of the variables in this investigation are stationary at either $I(0)$ or $I(1)$, as shown by the ADF unit root test. With these variable attributes, it is advised to conduct further research on the ARDL model, which was initially introduced by Pesaran et al. (1999). Later, the study used the Schwarz Information Criterion (SIC), Akaike Information Criterion (AIC) criteria and Hannan–Quinn Information Criterion (HQ) to determine the best lag selection. The investigation depicts one lag for dependent variable and two lag for independent as suggested by above three criteria for all three economies, and Table 4.1 shows the long-run analysis and bound test results.

Results of Long-Run Form and Bound Test

The existence of a long-term relationship is a crucial sign that further ARDL testing should be conducted. The ARDL methods can be applied to long-run estimations if a long-run relationship exists. The long-run form and bound test verified the long-run association's existence. In the lower portion of Table 4.1, the computed F-test values are 7.473672 and 6.281910 for China and the United States, respectively, which are greater than the upper bound's $I(1)$ value of 3.77 and 4.35 at the 10% and 5% level of significance, respectively. Consequently, it is demonstrated that the long-run agriculture productivity of China and the United States is influenced by the studied variable. In the case of India, the F-test value is 1.06187, which is less than the upper bound and lower bound values. This means that in the long run, there is no association between agriculture productivity and FDI inflows, CO2 emissions or forest coverage. However, the study still needs to

Table 4.1. Long-Run Analysis and Bound Test Results.

Long-Run Analysis Results of ARDL Model

Variables	India		China		United States	
	Coefficient [t-Statistics]	Prob.	Coefficient [t-Statistics]	Prob.	Coefficient [t-Statistics]	Prob.
lnVA_AGR(−1)	0.375236* [7 361990]	0.0000	0.754853* [11.81878]	0.0000	0.199623 [1.177731]	0.2509
lnFDI_AGR	0.023426*** [1.390187]	0.0704	0.049268*** [1.717641]	0.0969	0.015776 [0.933307]	0.3604
lnCO2E	0.043505 [0.388918]	0.7006	0.403306** [3.276797]	0.0028	1.389141** [3.297295]	0.0032
lnCO2E(−1)	–	–	–	–	−0.805405* [−1.771113]	0.0898
lnCO2E(−2)	–	–	–	–	– 1.040457** [−2.185821]	0.0393
lnFC	8.881759 [0.286475]	0.7769	0.025724 [0.035129]	0.9722	11.61449 [0.257153]	0.7994
lnFC(−1)	−45.59082 [−0.376783]	0.3890	–	–	76.37874 [0.956821]	0.3486
lnFC(−2)	37.53099 [1.490091]	0.1487	–	–	−63.96982 [−1.508416]	0.1451

(Continued)

Table 4.1. *(Continued)*

Long-Run Analysis Results of ARDL Model

Variables	India		China		United States	
	Coefficient [*t*-Statistics]	Prob.	Coefficient [*t*-Statistics]	Prob.	Coefficient [*t*-Statistics]	Prob.
C	−10.31141 [−0.44899]	0.6578	−3.802693 [−0.426827]	0.6728	− 342.4216** [−4.110926]	0.0004

Long Run Form and Bound Test Result

Bounds Critical Values			*F*-Calculated Value		
Significance	Lower Bound *I*(0)	Upper Bound *I*(1)	India	China	USA
10%	2.72	3.77	1.06187	7.473672	6.281910
5%	3.23	4.35			
2.5%	3.69	4.89			
1%	4.29	5.61			

Source: Author estimation using Eviews 10.

Note: ***, ** and * shows the significance level of coefficient at 1%, 5% and 10%, respectively.

evaluate the model's long-run coefficient. The bound test technique is used to get the F statistic value. The study used upper and lower critical bounds proposed by Banerjee et al. (1993) to compare and assess the significance level to verify the presence of associations in the long run. The presence of the long-run association can also be confirmed by looking at the significant and negative values of ECM.

Results of Long-Run Analysis

Table 4.1 indicates the long-run results: FDI inflows and agriculture productivity in lag period one was found to be statistically significant at 10% and 5% levels to stimulate the current period agriculture productivity of India and China measured by value-added to the economy. These results are in line with those of earlier research by researchers like Oloyede (2014) and Ali and Asgher (2016). They also concluded that FDI is essential to raising, maintaining and expanding the agricultural sector's output in the corresponding developing nations. While agriculture productivity in China and the United States was highly influenced by CO2 emissions, CO2 emissions had no statistically significant impact on Indian agriculture productivity. The negative and significant coefficient of lag in CO2 emissions showed that the rise of carbon emissions in the United States hurts agricultural productivity.

A decrease in agricultural production leads to a reduction in the growth of the agriculture sector. In the case of China, the effect of CO2 emissions on agriculture productivity is positive and significant. The study also found that the agricultural productivity of the ICU economy does not depend on forest coverage. Numerous studies conducted in response to FDI on agricultural productivity revealed notable and beneficial benefits. For example, the results of Furtan and Holzman (2004) confirmed that United States FDI (USFDI) greatly enhances Canada's agricultural growth and food security. Furthermore, Ogbanje et al. (2010) confirmed that, in the instance of Nigeria, there was a robust and favourable correlation between FDI and the expansion of the agriculture sector.

Results of Short-Run Analysis

Table 4.2 represents the short-run coefficients of the model. The coefficient of the ECM of the selected economy is negatively significant, and the negative coefficient value of India (−0.1248), China (−0.2451) and the United States (−0.8004) signifies a convergence to the equilibrium path in the long run. The coefficient of ECM magnitude shows a short-term adjustment process. In the short run, CO2 emissions and forest coverage have the potential to significantly influence the agricultural productivity of the United States, as per empirical results.

Furthermore, the right side of Table 4.2 showed the diagnostic tests for the model. The R^2 value is 0.324545 (India), 0.516363 (China) and 0.645645 (USA), indicating that variation in the productivity of the agriculture sector is explained by the independent variables. The Durbin–Watson statistics of all three economies indicate the absence of correlation issues in our model. As per the results of

Table 4.2. Result of ECM (Short-Run) Regression, Diagnostic Tests and Causality Test.

ECM Regression Analysis Result							
	Variables	Coefficient [*t*-Statistics]	Prob.	Diagnostic tests			Prob.
INDIA	D(lnFC)	8.881780 [0.380076]	0.7071	R^2	0.324545	Serial Correlation(A)	2.223825 (0.3289)
INDIA	D(lnFC(−1))	−37.55101 [−1.665906]	0.1082	Adjusted R^2	0.252175	Normality Test(B)	8.592705 (0.013618)
INDIA	ECM(−1)	−0.124764** [−2.180496]	0.0388	Durbin-Watson Stat	2.166217	Heteroscedasticity(C)	11.52472 (0.4846)
CHINA	ECM(−1)	−0.245142** [−5.753059]	0.0000	R^2 Adjusted R^2 Durbin-Watson Stat	0.516363 0.500762 1.997937	Serial Correlation(A) Normality Test (B) Heteroscedasticity(C)	0.652 (0.721) 0.627 (0.730) 5.471 (0.242)
UNITED STATES	D(lnCO2E)	1.389141** [4.052586]	0.0005	R^2	0.645645	Serial Correlation (A)	3.002417 (0.2229)
UNITED STATES	D(lnCO2E(−1))	1.040457** [2.569140]	0.0172	Adjusted R^2	0.577499	Normality Test(B)	0.963397 (0.617733)
UNITED STATES	D(lnFC)	11.61447 [0.336346]	0.7397				
UNITED STATES	D(lnFC(−1))	63.96984* [1.836360]	0.0793	Durbin-Watson Stat	1.925455	Heteroscedasticity(C)	4.580166 (0.7110)
UNITED STATES	ECM(−1)	−0.800377** [−5.329649]	0.0000				

VEC Granger Causality/Block Exogeneity Wald Test Result									
	Excluded	Chi-Sq. Value	*P* value	Implication		Excluded	Chi-Sq. Value	*P* value	Implication
Dependent Variable: lnVA_AGR					Dependent Variable: lnFDI_AGR				
INDIA	lnFDI_AGR	1.468499	0.4799	No Causality	INDIA	lnVA_AGR	10.45238	0.0054	Causality exist
INDIA	lnCO2E	0.043293	0.9786	No Causality	INDIA	lnCO2E	1.427537	0.4898	No Causality
INDIA	lnFC	2.194933	0.3337	No Causality	INDIA	lnFC	0.243776	0.8852	No Causality
CHINA	lnFDI_AGR	1.685305	0.4306	No Causality	CHINA	lnVA_AGR	3.619004	0.1637	No Causality
CHINA	lnCO2E	13.20963	0.0014	Causality exist	CHINA	lnCO2E	3.114394	0.2107	No Causality
CHINA	lnFC	0.313682	0.8548	No Causality	CHINA	lnFC	2.660362	0.2644	No Causality
USA	lnFDI_AGR	0.137261	0.9337	No Causality	USA	lnVA_AGR	1.638275	0.4408	No Causality
USA	lnCO2E	2.628426	0.2687	No Causality	USA	lnCO2E	0.056390	0.9722	No Causality
USA	lnFC	16.76181	0.0002	Causality exist	USA	lnFC	1.581367	0.4535	No Causality
Dependent Variable: lnCO2E					Dependent Variable: lnFC				
INDIA	lnVA_AGR	2.005324	0.3669	No Causality	INDIA	lnVA_AGR	8.685370	0.0130	Causality exist
INDIA	lnFDI_AGR	0.962365	0.6181	No Causality	INDIA	lnFDI_AGR	0.642134	0.7254	No Causality
INDIA	lnFC	0.894254	0.6395	No Causality	INDIA	lnCO2E	3.586962	0.1664	No Causality
CHINA	lnVA_AGR	13.35032	0.0013	Causality exist	CHINA	lnVA_AGR	4.498935	0.1055	No Causality
CHINA	lnFDI_AGR	4.091223	0.1293	No Causality	CHINA	lnFDI_AGR	0.351263	0.8389	No Causality
CHINA	lnFC	8.414872	0.0149	Causality exist	CHINA	lnCO2E	6.122144	0.0468	Causality exist
USA	lnVA_AGR	1.269277	0.5301	No Causality	USA	lnVA_AGR	2.857640	0.2396	No Causality
USA	lnFDI_AGR	1.180282	0.5542	No Causality	USA	lnFDI_AGR	17.54318	0.0002	Causality exist
USA	lnFC	1.627614	0.4432	No Causality	USA	lnCO2E	9.690996	0.0079	Causality exist

Source: The author's calculation.

Notes: (A) Based on Lagrange multiplier test of residual serial correlation (Breusch–Godfrey Serial Correlation LM Test).

(B) Based on a test of skewness and kurtosis of residuals (Jarque–Bera test of normality).

(C) Based on the white heteroscedasticity test with no cross terms yields (Breusch–Pagan–Godfrey Test). Respective probability values are presented in parentheses.

the diagnostics test, there is no indication of serial correlation or heteroscedasticity, and according to the Jarque–Bera test, the residuals are normally distributed in the model except in India. Furthermore, the absence of normalcy does not automatically imply that the model being employed is flawed; rather, it may point to additional variables that could adequately account for the relationship pattern (Geamănu, 2014).

Result From the Stability Test Graph

The study evaluates the long-term relationship among FDI inflows, agriculture productivity, CO2 emissions and forest coverage in the ICU economy. To ensure the stability of our model, the study relies on the 'CUSUM' and 'CUSUM' square tests. The stability of the coefficients during the estimation period is displayed in panels of Fig. 4.1. The CUSUM and CUSUM squares used to investigate the parameter stability are represented by the straight line, which displays the 5% critical boundaries for each. As the plots in both graphs fall between the key bounds showing the stability of the model's parameters, our model is structurally sound.

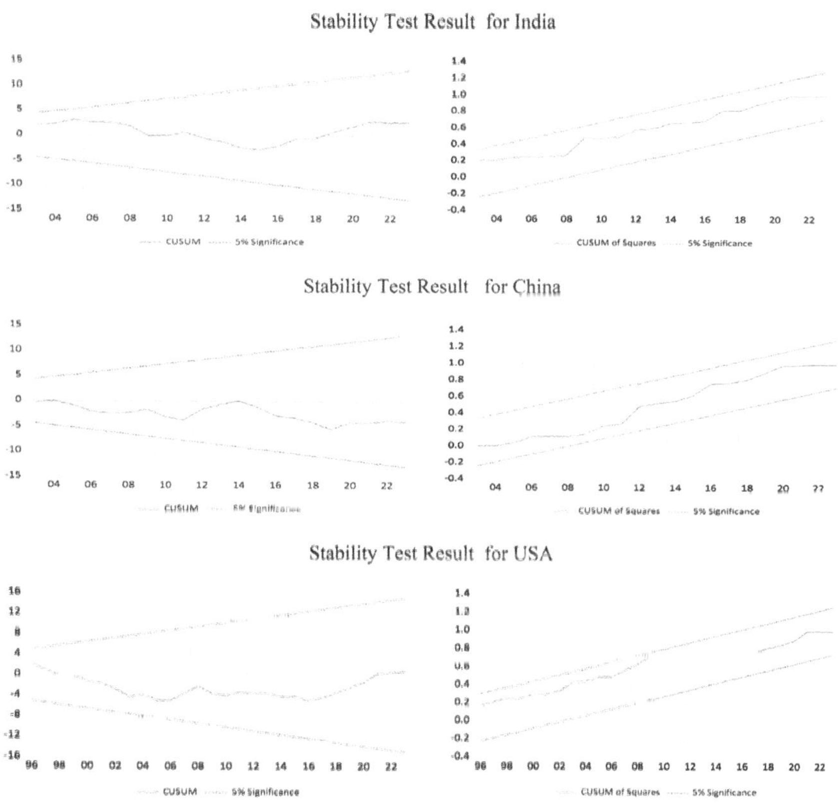

Fig. 4.1. Diagnostic Graphs for Stability Test. *Source:* The author.

Result From the Causality Test

The study considered the Granger causality test through vector error correction to estimate the direction of the long-run association among the variables. The result of the causality test is presented in lower portion of Table 4.2. An essential test for determining the cointegration relationship between the variables under analysis is the Granger causality test. In our investigation, we will thus examine the causal link between FC, $CO2$ emissions, FDI_AGR and VA_AGR. A long-term link between variables suggests that at least one direction of causality must exist (Granger, 1988).

As per the result of the VEC Granger causality test, there is a unidirectional causal relationship between agriculture productivity, FDI inflows in the agriculture sector and forest coverage in India. However, long-run form and bound tests suggest there is no long-run cointegration among agriculture productivity, FDI inflows, CO2 emissions and forest coverage in India. The result for China showed a bidirectional causal link between CO2 emissions and agricultural yield and forest coverage and CO2 emissions. However, in the case of the United States, a unidirectional causal association has been found from forest coverage to agricultural output and from FDI and CO2 emissions to forest coverage.

Conclusion and Recommendations

The primary objective of this empirical study was to investigate the role of FDI inflows, CO2 emissions and forest coverage from 1990 to 2023 on the agriculture productivity of the ICU economy in the short and long run. An empirical investigation revealed that in the long run, followed by the ARDL bound test, there was no association between Indian agriculture productivity, FDI inflows, CO2 emissions and forest area. However, FDI inflows have a short-run influence on agricultural productivity. The VEC Granger causality test suggests that a unidirectional causal link exists from agricultural productivity to FDI inflows and forest area. The decisive role of FDI inflows in agricultural activities proved that policymakers should take the necessary steps to boost investment in the sector. So the appropriate authority should ease the norms for foreign investors in the Indian agriculture sector to attract more FDI so that the sector can fill up the shortage of insufficient corpus.

In Chinese agriculture productivity, the study found a positive effect of CO2 emissions heavily in the long run, which means CO2 emission-based agriculture processes exist in China, whereas FDI inflows have a small positive impact on agricultural yield. The bidirectional causal effect between CO2 emission and agricultural yield and forest coverage and CO2 emission has been revealed by the VEC Granger causality test, which also indicates that CO2 emission-based agriculture and forestation processes are followed by China. Chinese policymakers should focus on reducing CO2 emissions and installing more environmentally friendly machinery in their agriculture and forestation processes. The policymakers also consider the ease of FDI norms in agriculture and the allied sectors.

The study of the United States revealed that in the long run, CO_2 emissions hurt US agriculture sector output, whereas in the short run, the positive impact of CO_2 emissions indicates the existence of a machinery-based agriculture system. In the short run, forest coverage also has a positive influence on agricultural productivity. A unidirectional causal effect has been established by the Granger causality test from forest coverage to agriculture productivity and from FDI inflows and CO_2 emissions to forest coverage. As agriculture productivity has been adversely affected by CO_2 emissions, policymakers should ensure the use of more environmentally friendly technology that increases forestation and reduces carbon emissions.

While the above analysis indicates interesting insights in ARDL bound testing and causality testing, it should be noted that the formulation of efficient environmental policies is likely to reduce carbon emissions and is necessary to increase forest area and agriculture productivity. The government should adopt specific policies to help the agriculture sector by providing environmentally friendly technology to improve agricultural productivity.

References

Alhassan, H. (2021). The effect of agricultural total factor productivity on environmental degradation in sub-Saharan Africa. *Scientific African*, *12*, e00740. https://doi.org/10.1016/j.sciaf.2021.e00740

Ali, H., & Asgher, M. T. (2016). The role of the sectoral composition of foreign direct investment on economic growth: A policy proposal for CPEC and regional partners. *Pakistan Development Review*, 89–103.

Angelsen, A., Jagger, P., Babigumira, R., Belcher, B., Hogarth, N. J., Bauch, S., Börner, J., Smith-Hall, C., & Wunder, S. (2014). Environmental income and rural livelihoods: A global-comparative analysis. *World Development*, *64*, S12–S28.

Assessment, M. E. (2001). *Millennium ecosystem assessment* (Vol. 2). Millennium Ecosystem Assessment. https://chapter.ser.org/europe/files/2012/08/Harris.pdf

Asumadu-Sarkodie, S., & Owusu, P. A. (2016). The relationship between carbon dioxide and agriculture in Ghana: A comparison of VECM and ARDL model. *Environmental Science and Pollution Research*, *23*, 10968–10982.

Balsalobre-Lorente, D., Driha, O. M., Bekun, F. V., & Osundina, O. A. (2019). Do agricultural activities induce carbon emissions? The BRICS experience. *Environmental Science and Pollution Research*, *26*, 25218–25234. https://link.springer.com/article/10.1007/s11356-019-05737-3

Banerjee, A., Dolado, J. J., Galbraith, J. W., & Hendry, D. (1993). *Co-integration, error correction, and the econometric analysis of non-stationary data*. Oxford University Press.

Ben Jebli, M., & Ben Youssef, S. (2017). Renewable energy consumption and agriculture: Evidence for cointegration and granger causality for Tunisian economy. *The International Journal of Sustainable Development and World Ecology*, *24*(2), 149–158. https://doi.org/10.1080/13504509.2016.1196467

Borlaug, N. (2007). Feeding a hungry world. *Science*, *318*(5849), 359.

Burney, J. A., Davis, S. J., & Lobell, D. B. (2010). Greenhouse gas mitigation by agricultural intensification. *Proceedings of the National Academy of Sciences, 107*(26), 12052–12057.

Carvalheiro, L. G., Seymour, C. L., Veldtman, R., & Nicolson, S. W. (2010). Pollination services decline with distance from natural habitat even in biodiversity-rich areas. *Journal of Applied Ecology, 47*(4), 810–820.

Chandio, A. A., Akram, W., Ahmad, F., & Ahmad, M. (2020). Dynamic relationship among agriculture-energy-forestry and carbon dioxide (CO2) emissions: Empirical evidence from China. *Environmental Science and Pollution Research, 27*, 34078–34089.

Chen, Y., Wang, Z., & Zhong, Z. (2019). CO2 emissions, economic growth, renewable and non-renewable energy production and foreign trade in China. *Renewable Energy, 131*, 208–216.

Chidiebere-Mark, N. M., Onyeneke, R. U., Uhuegbulem, I. J., Ankrah, D. A., Onyeneke, L. U., Anukam, B. N., & Chijioke-Okere, M. O. (2022). Agricultural production, renewable energy consumption, foreign direct investment, and carbon emissions: New evidence from Africa. *Atmosphere, 13*(12), 1981. https://doi.org/10.3390/atmos13121981

Clark, J. M. (1923). *Studies in the economics of overhead costs*. The University of Chicago Press.

Climate watch historical GHG emissions (1990–2020). (2023). World Resources Institute. https://www.climatewatchdata.org/ghg-emissions

Costanza, R., d'Arge, R., De Groot, R., Farber, S., Grasso, M., Hannon, B., Limburg, K., Naeem, S., O'neill, R. V., Paruelo, J., & Raskin, R. G. (1997). The value of the world's ecosystem services and natural capital. *Nature, 387*(6630), 253–260.

Costanza, R., De Groot, R., Sutton, P., Van der Ploeg, S., Anderson, S. J., Kubiszewski, I., Farber, S., & Turner, R. K. (2014). Changes in the global value of ecosystem services. *Global Environmental Change, 26*, 152–158.

Costanza, R., De Groot, R., Braat, L., Kubiszewski, I., Fioramonti, L., Sutton, P., Farber, S., & Grasso, M. (2017). Twenty years of ecosystem services: How far have we come and how far do we still need to go? *Ecosystem Services, 28*, 1–16.

Dickey, D. A., & Fuller, W. A. (1981). Likelihood ratio statistics for autoregressive time series with a unit root. *Econometrica: Journal of the Econometric Society*, 1057–1072.

Dogan, E., Taspinar, N., & Gokmenoglu, K. K. (2019). Determinants of ecological footprint in MINT countries. *Energy and Environment, 30*(6), 1065–1086.

Dogan, E., & Turkekul, B. (2016). CO2 emissions, real output, energy consumption, trade, urbanization and financial development: Testing the EKC hypothesis for the USA. *Environmental Science and Pollution Research, 23*, 1203–1213.

Engle, R. F., & Granger, C. W. (1987). Co-integration and error correction: Representation, estimation, and testing. *Econometrica: Journal of the Econometric Society*, 251–276.

FAO. (2016). *Greenhouse gas emissions from agriculture, forestry and other land use*. http://www.fao.org/3/a-i6340e.pdf

FAO. (2019). *The state of food and agriculture 2019. Moving forward on food loss and waste reduction*. Food and Agriculture Organization of the United Nations.

Food and Agriculture Organization of the United Nations Rome. (2022). https://openknowledge.fao.org/server/api/core/bitstreams/f6270800-eec7-498f-9887-6d937c4f575a/content

Furtan, W. H., & Holzman, J. J. (2004). *The effect of FDI on agriculture and food trade: An empirical analysis 1987–2001*. https://www150.statcan.gc.ca/n1/pub/21-601-m/21-601-m2004068-eng.pdf

Geamănu, M. (2014). VAR analysis on foreign direct investment in Romania. *Theoretical and Applied Economics*, *21*(4), 39–52.

Gill, A. R., Viswanathan, K. K., & Hassan, S. (2018). A test of environmental Kuznets curve (EKC) for carbon emission and potential of renewable energy to reduce green house gases (GHG) in Malaysia. *Environment, Development and Sustainability*, *20*, 1103–1114.

Gokmenoglu, K. K., & Taspinar, N. (2018). Testing the agriculture-induced EKC hypothesis: The case of Pakistan. *Environmental Science and Pollution Research*, *25*, 22829–22841.

Gokmenoglu, K. K., Taspinar, N., & Kaakeh, M. (2019). Agriculture-induced environmental Kuznets curve: The case of China. *Environmental Science and Pollution Research*, *26*, 37137–37151.

Granger, C. W. (1969). Investigating causal relations by econometric models and cross-spectral methods. *Econometrica: Journal of the Econometric Society*, 424–438.

Granger, C. W. (1988). Some recent development in a concept of causality. *Journal of Econometrics*, *39*(1–2), 199–211.

Grossman, G. M., & Krueger, A. B. (1991). *Environmental impacts of a North American free trade agreement*. https://doi.org/10.3386/w3914

Intergovernmental Panel on Climate Change. (2021). Summary for policymakers. In V. Masson-Delmotte, P. Zhai, A. Pirani, S. L. Connors, C. Péan, S. Berger, N. Caud, Y. Chen, L. Goldfarb, M. I. Gomis, M. Huang, K. Leitzell, E. Lonnoy, J. B. R. Matthews, T. K. Maycock, T. Waterfield, O. Yelekçi, R. Yu, & B. Zhou. (Eds.), *Climate change 2021: The physical science basis. Contribution of working group I to the sixth assessment report of the intergovernmental panel on climate change* (pp. 3–32). Cambridge University Press. https://www.ipcc.ch/report/ar6/wg1/downloads/report/IPCC_AR6_WGI_SPM.pdf. Accessed on February 23, 2024.

IPCC Working Group III. (2014). *Climate change 2014-mitigation of climate change. Synthesis report*. Intergovernmental Panel on Climate Change Cambridge University Press.

Keh, C. G., Tan, Y. T., Tang, S. E., Sim, J. J., & Lee, C. Y. (2023). Evaluating the role of forested area, agricultural land, energy consumption and foreign direct investment on Co2 emissions in Indonesia. *Journal of Tourism, Hospitality and Environment Management*, *8*(32), 72–87.

Khan, M. T. I., Ali, Q., & Ashfaq, M. (2018). The nexus between greenhouse gas emission, electricity production, renewable energy and agriculture in Pakistan. *Renewable Energy*, *118*, 437–451.

Kim, D. H., Sexton, J. O., & Townshend, J. R. (2015). Accelerated deforestation in the humid tropics from the 1990s to the 2000s. *Geophysical Research Letters*, *42*(9), 3495–3501.

Klein, A. M., Steffan-Dewenter, I., & Tscharntke, T. (2003). Pollination of Coffea canephora in relation to local and regional agroforestry management. *Journal of Applied Ecology*, *40*(5), 837–845.

Klein, A. M., Vaissière, B. E., Cane, J. H., Steffan-Dewenter, I., Cunningham, S. A., Kremen, C., & Tscharntke, T. (2007). Importance of pollinators in changing landscapes for world crops. *Proceedings of the Royal Society B: Biological Sciences, 274*(1608), 303–313.

L'Roe, J., & Naughton-Treves, L. (2014). Effects of a policy-induced income shock on forest-dependent households in the Peruvian Amazon. *Ecological Economics, 97*, 1–9.

Liu, X., Zhang, S., & Bae, J. (2017). The nexus of renewable energy-agriculture-environment in BRICS. *Applied Energy, 204*, 489–496.

Lokonon, B. O., & Mounirou, I. (2019, November). Does foreign direct investment impede forest area in sub-Saharan Africa? In *Natural resources forum* (Vol. 43, No. 4, pp. 230–240). Blackwell Publishing Ltd.

Mäder, P., Fliessbach, A., Dubois, D., Gunst, L., Fried, P., & Niggli, U. (2002). Soil fertility and biodiversity in organic farming. *Science, 296*(5573), 1694–1697.

Ogbanje, E. C., Okwu, O. J., & Saror, S. F. (2010). An analysis of foreign direct investment in Nigeria: The fate of Nigeria's agricultural sector. *Production Agriculture and Technology Journal, 6*(2), 15–25.

Olanipekun, I. O., Olasehinde-Williams, G. O., & Alao, R. O. (2019). Agriculture and environmental degradation in Africa: The role of income. *Science of the Total Environment, 692*, 60–67.

Oloyede, B. B. (2014). Impact of foreign direct investment on agricultural sector development in Nigeria, (1981–2012). *Arabian Journal of Business and Management Review, 3*(12), 14–24.

Panayotou, T. (1993). *Empirical tests and policy analysis of environmental degradation at different stages of economic development*. https://d1wqtxts1xzle7.cloudfront.net/79436985/93B09_31_engl-libre.pdf?1642973134=&response-content-disposition=inline%3B+filename%3DEmpirical_tests_and_policy_analysis_of_e.pdf&Expires=1724349700&Signature=Z6XTcgYJo~wMpls1OYZ24VEsvzMJnQQBEMyLyN1sTJvtMnZFZ0~NSSwsFUeWhcr0t4Vqo1~3uxKRJ5dMG2G62MH8gk5QbyyDlpe2RQbFj2r-IRsCZ5IIGpOYXJuLxttNXqDYkSEN99~drL5CoUmkt-OEpsO~gl5b6dnFlDzqBREBfWWanOSp~GKnpD8kyjMQRnEx3qwcAIkvl4mJeOfOStyxDE4w3yGpUYgMPuG3DInaCGhG0qrRKQsmSOb0-IqM9IjwAD6zeNUm6wcOwPGecu7yg7awoz1B5Bf7ssJyTt6lUKlf~ikZGXZ17mtsl39ANwgu9wr4blRfEDkQ38QrrQ__&Key-Pair-Id=APKAJLOHF5GGSLRBV4ZA

Pata, U. K. (2021). Linking renewable energy, globalization, agriculture, CO2 emissions and ecological footprint in BRIC countries: A sustainability perspective. *Renewable Energy, 173*, 197–208.

Pata, U. K., & Aydin, M. (2020). Testing the EKC hypothesis for the top six hydropower energy-consuming countries: Evidence from fourier bootstrap ARDL procedure. *Journal of Cleaner Production, 264*, 121699. https://doi.org/10.1016/j.jclepro.2020.121699

Pellegrini, G., Ingrao, C., Camposeo, S., Tricase, C., Conto, F., & Huisingh, D. (2016). Application of water footprint to olive growing systems in the Apulia region: A comparative assessment. *Journal of Cleaner Production, 112*, 2407–2418.

Pesaran, M. H., Shin, Y., & Smith, R. P. (1999). Pooled mean group estimation of dynamic heterogeneous panels. *Journal of the American Statistical Association, 94*(446), 621–634.

Pesaran, M. H., Shin, Y., & Smith, R. J. (2001). Bounds testing approaches to the analysis of level relationships. *Journal of Applied Econometrics, 16*(3), 289–326.

Prastiyo, S. E., Irham, Hardyastuti, S., & Jamhari, F. (2020). How agriculture, manufacture, and urbanization induced carbon emission? The case of Indonesia. *Environmental Science and Pollution Research*, *27*(33), 42092–42103.

Rafiq, S., Salim, R., & Apergis, N. (2016). Agriculture, trade openness and emissions: An empirical analysis and policy options. *The Australian Journal of Agricultural and Resource Economics*, *60*(3), 348–365.

Ramachandra, T. V., Aithal, B. H., & Sreejith, K. (2015). GHG footprint of major cities in India. *Renewable and Sustainable Energy Reviews*, *44*, 473–495.

Raudsepp-Hearne, C., Peterson, G. D., & Bennett, E. M. (2010). Ecosystem service bundles for analyzing tradeoffs in diverse landscapes. *Proceedings of the National Academy of Sciences*, *107*(11), 5242–5247.

Rehman, A., Ozturk, I., & Zhang, D. (2019). The causal connection between CO2 emissions and agricultural productivity in Pakistan: Empirical evidence from an autoregressive distributed lag bounds testing approach. *Applied Sciences*, *9*(8), 1692. https://doi.org/10.3390/app9081692

Ricketts, T. H., Daily, G. C., Ehrlich, P. R., & Michener, C. D. (2004). Economic value of tropical forest to coffee production. *Proceedings of the National Academy of Sciences*, *101*(34), 12579–12582.

Ridzuan, N. H. A. M., Marwan, N. F., Khalid, N., Ali, M. H., & Tseng, M. L. (2020). Effects of agriculture, renewable energy, and economic growth on carbon dioxide emissions: Evidence of the environmental Kuznets curve. *Resources, Conservation and Recycling*, *160*, 104879. https://doi.org/10.1016/j.resconrec.2020.104879

Romer, P. M. (1986). Increasing returns and long-run growth. *Journal of Political Economy*, *94*(5), 1002–1037.

Seker, F., Ertugrul, H. M., & Cetin, M. (2015). The impact of foreign direct investment on environmental quality: A bounds testing and causality analysis for Turkey. *Renewable and Sustainable Energy Reviews*, *52*, 347–356.

Selvanathan, S., Jayasinghe, M. S., Selvanathan, E. A., Abbas, S. A., & Iftekhar, M. S. (2023). Energy consumption, agriculture, forestation and CO2 emission nexus: An application to OECD countries. *Applied Economics*, *55*(37), 4359–4376.

Shafik, N., & Bandyopadhyay, S. (1992). *Economic growth and environmental quality: Time-series and cross-country evidence* (Vol. 904). World Bank Publications.

Shahbaz, M., Khan, S., Ali, A., & Bhattacharya, M. (2017). The impact of globalization on CO2 emissions in China. *Singapore Economic Review*, *62*(04), 929–957.

Sloan, S., & Sayer, J. A. (2015). Forest resources assessment of 2015 shows positive global trends but forest loss and degradation persist in poor tropical countries. *Forest Ecology and Management*, *352*, 134–145.

Solow, R. M. (1956). A contribution to the theory of economic growth. *Quarterly Journal of Economics*, *70*(1), 65–94.

Tariq, M., Xu, Y., Muhammad, F., & Alam, K. M. (2019). The dirty energy dilemma via financial development and economic globalization in Pakistan: New evidence from asymmetric dynamic effects. *Environmental Science and Pollution Research*, *26*, 25500–25512.

TEEB, R. O. (2010). *Mainstreaming the economics of nature*. TEEB. https://www.iges.or.jp/jp/publication_documents/pub/translation/jp/13140/teeb_synthesis_j.pdf

Timmer, C. P. (2010). Reflections on food crises past. *Food Policy*, *35*(1), 1–11.

Timperley, J. (2019). *The carbon brief profile*. https://www.carbonbrief.org/the-carbon-brief-profile-india/#:~:text=Jocelyn%20Timperley,-14.03.2019%20%7C%204=India%20is%20the%20world's%20third,after%20China%20and%20the%20US=Coal%20power%20plants%2C%20rice%20paddies,well%20below%20the%20global%20average

Udemba, E. N. (2020). A sustainable study of economic growth and development amidst ecological footprint: New insight from Nigerian perspective. *Science of the Total Environment*, *732*, 139270. https://doi.org/10.1016/j.scitotenv.2020.139270

United Nations. (2022). *World population prospects: The 2022 version*. https://population.un.org/wpp/Publications/Files/WPP2022_Highlights.pdf. Accessed on February 24, 2022.

Vedeld, P., Angelsen, A., Bojö, J., Sjaastad, E., & Berg, G. K. (2007). Forest environmental incomes and the rural poor. *Forest Policy and Economics*, *9*(7), 869–879.

Wang, L., Vo, X. V., Shahbaz, M., & Ak, A. (2020). Globalization and carbon emissions: Is there any role of agriculture value-added, financial development, and natural resource rent in the aftermath of COP21? *Journal of Environmental Management*, *268*, 110712. https://doi.org/10.1016/j.jenvman.2020.110712

World Development Report (2008). *Agriculture for development*. The World Bank. https://openknowledge.worldbank.org/entities/publication/ba969388-b5eb-5155-b8f2-6d323a6e5a52

Yasmeen, R., Padda, I. U. H., Yao, X., Shah, W. U. H., & Hafeez, M. (2021). Agriculture, forestry, and environmental sustainability: The role of institutions. *Environment, Development and Sustainability*, 1–25.

Yilanci, V., & Pata, U. K. (2020). Investigating the EKC hypothesis for China: The role of economic complexity on ecological footprint. *Environmental Science and Pollution Research*, *27*(26), 32683–32694.

Zafeiriou, E., & Azam, M. (2017). CO2 emissions and economic performance in EU agriculture: Some evidence from Mediterranean countries. *Ecological Indicators*, *81*, 104–114.

Chapter 5

The Agricultural Exports on the Renegotiation of USCM: The Case of Sinaloa

José G. Vargas-Hernández[a], *Elsa Patricia Orozco Quijano*[b] *and M. C. Omar C. Vargas-González*[c]

[a]Tecnológico Nacional de México, ITSF, México
[b]Laurentian University, Canada
[c]Instituto Tecnológico de México, Campus Cd. Guzmán, México

Abstract

Like many other countries, Mexico has invested in finding measures to attract foreign direct investment (FDI) to its territories. In 1994, Mexico signed the United States–Canada–Mexico (USCM), a pact that positions Mexico as the United States' greatest direct exporter, with the prospect of leaving the treaty following the renegotiating of the USCM. This study seeks to analyse the benefits and drawbacks of renegotiation inside the USCM. It is said that the renegotiation of the Treat of Mexico, United States and Canada (Tratado de México, Estados Unidos y Canadá, TMEC) does not have to cause a wave of anxiety surrounding agricultural commodities and goods from the State of Sinaloa's exports because the renegotiation was accepted and a fair arrangement was established for the three countries. Looking directly at Sinaloa's agricultural exports, the study suggests that such a renegotiation is not beneficial for Sinaloan agricultural exportation putting forth the evidences to support the hypothesis applied.

Keywords: Renegotiation; NAFTA; Sinaloa exports; FDI; agroeconomy; USCM

Growth, Poverty and Developmental Aspects of Agriculture, 63–78

doi:10.1108/978-1-83608-076-320241005

Introduction

In Eduardo Galeano's book 'The Open Veins of Latin America', he stipulates the phrase: The people who buy rule, the people who sell served; trade must be balanced to ensure freedom; the people who want to die sell to only one people, and the people who want to save sell to more than one (Galeano, 1970).

The North American Free Trade Agreement (NAFTA), agreed in 1992 by the governments of Mexico, Canada and the United States, went into effect in 1994 to create a free trade area with lower costs for the flow of commodities between the three nations. The NAFTA, which is currently known as TMEC, came into effect in 1994. Since 1994, the NAFTA has created one of the world's largest free trade zones, laying the groundwork for strong economic growth and greater prosperity in Canada, the United States and Mexico.

For 15 years, the NAFTA has proved how free trade promotes wealth and competitiveness by benefiting families, farmers, workers, manufacturers and consumers (NAFTA de hoy, 2014). The NAFTA is a regional agreement between the Canadian, Mexican and American governments to establish a free trade zone (Secretariado del TLCAN, 2017).

The NAFTA, or United States, Canada and Mexico Agreement, is a system of rules under which Mexico, the United States and Canada agree to trade and acquire goods and services (Castro, 2008). It was an innovative pact designed to open and grow the North American market. Since then, the NAFTA has gradually removed the majority of tariff and non-tariff barriers to trade and investment between Canada, the United States and Mexico, resulting in the formation of a stable and confident framework for long-term investments.

The Canada–United States Free Trade Agreement (FTA) predates the TMEC. President Donald Trump was in a renegotiation crisis as a result of the arrival into power of the US government, which promised to rearrange the regulations set in the treaty with faithful benefits in his favour, and if not, he announced his country's withdrawal from the deal. The renegotiation has already been completed, but there is widespread concern in Mexico about the impact on its exports, as it is the primary provider of raw materials and a range of merchandise to the United States.

Currently, a number of changes have occurred in the decisions of the governments involved in the TMEC, namely the United States. As a result, it is critical to conduct a thorough analysis of the impact of the renegotiation on Sinaloa exports, as the state of Sinaloa has gained the nickname 'Mexico's breadbasket' due to the abundance of raw materials produced. It is the leading state in selling agricultural products to its neighbouring country (the United States). For the reasons mentioned, the current study is spatially constrained in order to establish a firm foundation for analysing the benefits and drawbacks of the TMEC renegotiation in agricultural exports Sinaloans.

As a result, it is critical to investigate and determine the impact of the TMEC renegotiation on Sinaloan agricultural exports, as the state known as 'the granary of Mexico' for its wealth in agricultural crops has earned the first place in the extraction of these raw materials to be exported to the neighbouring country, the United States.

That is why the purpose of this research is to determine the advantages and disadvantages of the production, trading and distribution networks that will dominate following the renegotiation. Captivating as indicators, Mexico's exports and trade, the GDP, GDP per capita and transportation logistics for export shipments, this chapter intends to:

- Analyse the renegotiation of the TMEC.
- Determine the advantages of Sinaloan exports.
- Determine the disadvantages of Sinaloan exports.
- Analyse the productions of agriculture in Sinaloa.
- Analyse Sinaloa's trade in relation to the United States.

Some of the concerns addressed in this study include: Will the renegotiation of the TMEC have a detrimental influence on Sinaloan agricultural exports? What might Sinaloa's agricultural export production look like if the TMEC was renegotiated? What impact would the TMEC's renegotiation advantages have on agricultural export commerce in Sinaloa? What might Sinaloa's agricultural export distribution network look like if the TMEC were renegotiated?

Mexico, like other nations, has engaged in efforts to encourage foreign direct investment in its territories. As a result, in 1994, it signed the NAFTA pact with the United States of America and Canada, which established Mexico as the largest direct exporter to the United States, a country that may exit the treaty by renegotiating the TMEC. As a result, it is critical to conduct this research to determine the benefits and drawbacks of the above-mentioned implementation in Sinaloense agricultural exports, which are one of the main export productions and a source of employment for many Sinaloenses. It is expected that if the renegotiation was not feasible for our country, Mexico, it would have to abandon the treaty.

The primary goal of the study is to analyse the benefits and drawbacks of renegotiating the TMEC based on Sinaloa's agricultural exports. The specific objectives are to determine production in Sinaloa's agricultural exports based on TMEC renegotiation, trade in agricultural exports from Sinaloa based on TMEC renegotiation advantages and distribution network in Sinaloa's agricultural exports based on TMEC renegotiation disadvantages.

The Impact of the Agroeconomy Sector of Sinaloa in Mexico

This chapter reviews the influence of the renegotiation of TEMEC on agricultural exports from the state of Sinaloa, which is expected to benefit from the change in the country's orientation abroad, given its traditional specialisation in vegetables for export to the market. In order to have reference parameters, Sinaloa is approached in the broader context of Mexico, while reflecting on basic aspects of its economic structure and the dynamics of agriculture, showing that it is still very important in the economic dynamics of the state of Sinaloa, but which, at the same time, is in other sectors of activity where the key is to attack the problem of low growth (CEDRSSA, 2020).

During several 6-year terms of government, he seeks to promote structural reforms that benefit the inefficient economy through exports. For example, a crusade against hunger is generated when, incongruously, the majority of the population that produces food is in a situation of extreme poverty. The lack of importance given to the agricultural sector is serious without highlighting the role that this sector theatres in the Mexican economy.

Mexico has had a production increase of 82.4% in 25 years, in 2019 the sector closed with a production of 282.5 million metric tons, a result 1.4% better than the previous year, although a decrease of the 25% by 2020, this derived from some drought problems suffered, mainly in beans and sugar cane. And in the case of the state of Sinaloa there was a lack of water for the dams. Likewise, in Sinaloa, a production of 5.5 million tons of white corn was achieved in the recent season (Codesín, 2021).

The State of Sinaloa in México, is the most important in agricultura production of grains in Mexico, which has earned it the nickname of 'the granary of Mexico', since grains such as corn, beans and chickpea are produced that occupy the first national places in as for the value of production. Likewise, the state stands out as the main producer and exporter of vegetables, such as red tomato, chili peppers and cucumber. One out of every two tomatoes that Mexico exports is from Sinaloa (Reyes Rodríguez & Quintero Reyes, 2020).

In the first semester of 2019, Sinaloa managed to export 2,414.03 million dollars in products from the agri-food, agriculture, agribusiness, manufacturing, fishing and mining sectors, against 2 thousand 139.86 million dollars in the same period of 2018, with which registered a growth of 12.8% in exports. The main export destinations were: the United States, Canada, Japan, Spain, South Korea, Venezuela and the Netherlands. Likewise, the main exportable products are: tomatoes, meat and beef cuts with and without bone, mango, bell pepper, beef viscera, fruit juices, frozen tuna and fillets, chickpeas, peppers, grapefruit oils, tapas and aluminium cans, plastics, corn, wheat, fish meal and sound equipment (Toledo, 2020).

The dynamics of the agricultural GDP of Sinaloa and Mexico tend to be very similar throughout the period, highlighting only that in the case of the agricultural GDP of Sinaloa, the variations are more abrupt, but in general, the dynamics are similar. Although both the agricultural GDP of the Sinaloa river and the national GDP have had periods of stagnation, the rate of growth continues to rise. Regarding this, despite the weakening of the policy aimed at the primary sector, the growth of agricultural productivity has been notable thanks to the strengthening of production chains and contract farming (Bruno et al., 2017).

Theoretical-Empirical Review of the Literature

David Ricardo's fundamental free trade theory states that participating in a free trade zone benefits all countries. In theory, the idea that each country should specialise based on its comparative advantages makes sense (Dillon, 1991). The TMEC promotes economic growth and dynamic commerce, encourages investment while also forming productive alliances, and caters to small and medium-sized businesses in a fair and predictable environment. TMEC partners enhance environmental conservation and increase economic prospects in North America (Audley et al., 2003).

After Wilbur Ross, Secretary of Commerce of the United States, declared 'a sensible negotiation' of the NAFTA, experts observe a measure in that country's position (El Financiero, 2017). Wilbur Ross, Secretary of Commerce of the United States, appears to have given the TMEC a break after saying in an interview with the NBC network that a 'sensible' reform of the pact would improve its weight, analysts concurred. The statements demonstrate that the 'TMEC is alive and capable of being modernised', he assured. The decision to strengthen the Mexican peso may have the goal of preventing Mexico from gaining an unfair advantage and reducing currency appreciation. This, in turn, implies that the United States wishes to continue dealing with the country, O'Farril explained. He added that the third course of action is to find measures to enhance wages and improve the living conditions of Mexican workers. However, Shrik cautioned that while it may appear that the balance is now tilting more in favour of maintaining the TMEC, what a cabinet member says is not necessarily the same as what Trump believes because uncertainty in Trump's relationship with Mexico has been a constant in his little more than 40 days in office. 'We don't know who we're going to wake up with tomorrow', Shrik remarked, alluding to Trump's unpredictable nature. Arturo Pérez Behr, president of the National Association of Importers and Exporters of the Mexican Republic, expressed concern prior to Trump's remarks (El Financiero, 2017).

Internal aid cannot be negotiated in a bilateral FTA because it is hard to determine which products enter a country with aid and which do not. Domestic aid has thus not been negotiated in this deal with the United States (Mypimes, 2017). There is now widespread agreement on the benefits of nation-state economic integration. Free trade is beneficial because it provides access to a greater number of commodities at lower prices, hence increasing societal welfare. It provides access to lower cost inputs and capital goods, potentially increasing the competitiveness of value-adding production sectors.

Free trade encourages efficient resource allocation because price signals are not distorted: when a good is in short supply, its price rises, signalling producers to increase supply (and consumers to reduce demand); when a good is abundant, its price falls, signalling producers to reduce supply (and consumers to increase demand) (Mypimes, 2017).

Agricultural Free Trade: Theory and Reality

From a macroeconomic standpoint, expanding the economy's current account means opening it up to the rest of the world's savings. However, the reality of international commerce, particularly in the agriculture sector, differs significantly from the first best theoretician's definition of free trade. Why? To answer this question, we must first recognise that the agricultural sector is unique and distinct from other sectors of the economy, and that preserving agricultural operations benefits society as a whole. When a product is acquired from a peasant, not only is the product obtained but also the legal and peaceful possession of national rural territory (Mypimes, 2017).

In practice, international trade in the agricultural sector hurts from three sorts of distortions or barricades – *First*, there are tariff barriers. These include ad valorem tariffs, special tariffs and quotas. *Second*, there are non-tariff obstacles, which can have more restrictive impacts on trade than tariff barriers. Among these forms of distortions, the sanitary technical and phyto/zoo requirements required for product importation stand out. *Third*, we have export subsidies. Given their financial resources, rich countries choose to protect the rural sector through domestic assistance and export subsidies.

An export is the shipment of national or nationalised goods for use or consumption overseas (Minister of Foreign Affairs and Cooperation, 2017). To protect our rural producers, developing countries must impose tariffs. It is critical to remember that border support (tariffs) is the focus of an FTA negotiation. The reasoning is straightforward. Internal aid cannot be negotiated in a bilateral FTA because it is hard to determine which products enter a country with aid and which do not. Domestic aid is not going to be negotiated in this pact with the United States (Mypimes, 2017).

Contextual Framework of Exports in Sinaloa

According to the Secretary of Economic Developement (Secretaría de Desarrollo Económico, SEDECO) and Coordinator of Promotion of Foreign Trade of the State of Guanajuato (Coordinadora de Fomento al Comercio Exterior del Estado de Guanajuato, COFOCE) information system, the agro-food and agriculture sectors accounted for 62% of total Sinaloan exports at the end of 2015. According to this source, the top exporting firm in Sinaloa is Sukarne SA de CV Grupo VIZ, which accounts for over 28% of total exports. The auto parts firm Deplhi de México SA de CV and Envases Universales de México are located in Los Mochis, Mazatlán, while the agro-industrial sector Citrofrut and Conservas La Costeña have activities in Rosario and Guasave. At the end of 2015, this system had 607 registered exporting enterprises (CODESIN, 2017).

In 2016, 14 nations opened their markets to receive animal products from Mexico, some for the first time and others with particular restrictions. The Secretariat of Agriculture, Livestock, Rural Development, Fisheries and Food (Secretaría de Agricultura, Ganadería, Desarrollo Rural, Pesca y Alimentación, SAGARPA) negotiations of sanitary protocols with the authorities of 14 nations and the European Union resulted in the opening of markets in 2016 for 18 Mexican products, nine of animal origin and nine of vegetables, according to the federal agency. The prohibition on exporting pathogen-free eggs and egg products to the European Union was lifted as a result of negotiations with various nationals conducted by the National Service of Health, Safety, and Agro-Food Quality (Senasica), and the market for Mexican honey reopened in Saudi Arabia.

In addition to sorghum bushes for chili, stevia plants for Guatemala, maize seed for Ecuador, chilli seed for Portugal and amaranth grain for consumption in neighbouring northern Turkey and the European Union. According to SAGARPA, this year, it is intended to expand the supply of Mexican products overseas, with discussions already underway with health services from more than 15 countries. About 13 Mexican-made products are scheduled to 'step on' the land of nations such as Russia, China, Indonesia, Iran, Singapore, Vietnam, Taiwan and Japan for the first time (Torres, 2017).

Materials and Methods

It is exploratory research, and the primary goal is to promote a deeper penetration and comprehension of the subject at hand. It is analytical because it is descriptive and is more closely related to statistical and control data in order to produce hypotheses about what happened or will happen, as well as forecast failures or events. And of the experimental variety, as it is based on experimentation or observations.

Research Design

Table 5.1 presents the research design to be used in the present study.

Research Instrument

The method employed in this research project is the bibliographic review analysis and hemerographic analysis because it emphasises on the renegotiation of the NAFTA, for which this research is mainly based on agricultural exports of the state of Sinaloa was suggested with indicators such as the US exit from the TMEC, exports-trade, GDP and GDP per capita of Mexico, opening to new markets, logistics. Materials and methods should be given in enough detail that others can reproduce and expand on published findings.

Data Analysis

According to preliminary data from the National Institute of Statistics and Geography (INEGI), the Mexican economy increased 2.9% on an annual basis in Q1 2016. This improvement reflects Mexico's greatest first quarter since 2013, as well as the strongest quarterly growth since the second quarter of 2014. While primary economic activities, including agriculture, had increased annually by 3% on gross domestic product, outpacing the 2.9% increase seen last year.

GDP per inhabitant is the economic worth of a nation's commodities and services that would be allocated to each inhabitant if wealth were divided evenly (BANXICO, 2017; CONAPO, 2017). The GDP is the total worth of a country's final products and services generated during a specific time period. GDP per capita, also known as per capita, is the same number divided by a country's total population. Mexico's GDP per capita hit an all-time high in 2014, yet only 2 years later, it fell by more than 20%.

Table 5.1. Research Design.

Variable	Description	Dimensions	Indicators	Research Instruments	Operationalisation of Variables	Statistical Analysis
X0 TMEC	'The North American Free Trade Agreement or NAFTA is a trade agreement between the three countries in North America: Canada, the United States of America and Mexico'.	Renegotiation Advantage Disadvantages	US exit from the treaty Exports Commerce	Bibliographic analysis Hemerographic analysis	Review of practical case Bibliographic analysis	Analytical study Exploratory study Empirical study
Y0 Agricultural exports	Product of the primary sector that you dare from cultivation as a production activity obtains raw materials	Production Commerce Red de distribution	GDP distribution network GDP per capita in Sinaloa Opening to new markets Logistics	Bibliographic analysis Hemerographic analysis	Review of practical case Hemerographic analysis	Analytical study Exploratory study Empirical study

Source: Own elaboration.

GDP per capita indicates the economic worth of a nation's goods and services that would be allocated to each resident if wealth were divided evenly. This suggests that if a country's GDP increases while its population remains steady, the GDP per capita rises; but if the population rises while the GDP remains constant or falls, the GDP per capita falls as well.

Export Logistics

Without an entrance door that allows for the correct movement of products and services, any country or region's international trade cannot succeed. Veracruz, Mexico's port, represents the country's economic and industrial progress from the 16th century. Today, its importance is undeniable, and its problems are becoming more complicated as industrial processes and product types evolve, putting any logistics system to the test. The logistics chain begins with raw material manufacturing and concludes in consuming centres. In this respect, seaports are nodes in the physical network of marine transport, and their competitiveness is determined by the speed, flexibility and safety of the services they provide for international trade and shipping.

According to an Inter-American Development Bank (IDB) study on cargo logistics in Latin America and the Caribbean, there are at least three viewpoints to consider when determining the importance of the logistics procedure in an economy: the generator cargo (industry), the cargo operator (services) and public policy. The ordered interaction of these three aspects allows for steady reductions in logistics operating expenses throughout time.

The Federal Government will invest 16,000 million pesos in the first stage of the expansion, and by the project's finish in 2030, they will have invested 70,000 million pesos, with 42,000 private and 28,000 public. The World Bank (Banco Mundial, 2010) placed Mexico 50th out of 155 nations in 2010, with a performance score of 3,051. Some consultants estimate that logistics expenses in Mexico account for 15.3% of GDP. Mexican enterprises' logistics expenditures account for 10.3% of sales, with 40% covering transportation costs and 60% covering stocks, order processing, storage and transportation operations management planning (Market, Real Estate, 2017).

The federal government and the IDB recently conducted an investigation on the National System of Logistics Platforms (SNPL), which stated the system as consisting of 85 logistics platforms, with a focus on infrastructure and service investment to improve supply chain efficiency. In this perspective, consider the performance of Mexican ports in recent years. In 2010, commercial imports by sea were 75.2 million tonnes, while 122.2 million tonnes were shipped to all parts of the world. In 2013, 81.1 million tonnes were imported, and 129.1 million tonnes were exported (Market, Real Estate, 2017). This development was made possible by Mexico's commerce on the Gulf and Caribbean coasts (15 ports) and in the Pacific Ocean (20 ports), with Veracruz serving as the most representative of the manufacturing industry.

According to the Economic Commission for Latin America and the Caribbean, Veracruz placed 18th in the area for containerised port movements in 2010 and 16th in the first half of 2013. On the other side, according to the IDB's 2014 port technical efficiency report, the port of Veracruz ranks 22nd out of 63 ports in Latin America and the Caribbean (Market, Real Estate, 2017). The port of Veracruz is the country's most significant commercial port. It handles 100% of purely commercial freight transits and is the only company that manages the six most essential cargo segments across the nation.

Exports decreased by 10.7%, moving 4,730,856 tonnes. Trade has prioritised port expansion since it would increase jobs and strengthen the regional and state economies. The majority of these automobiles originate from Puebla, Estado de México, Aguascalientes, Coahuila, Tamaulipas, Nuevo León and Veracruz. This port is the most important for the export and import of cars. According to the table below, Veracruz port handles 66% of total car exports via sea. In the future decades, the port of Veracruz will be one of the most significant logistical and marine enclaves in the country and on the Atlantic coast, connecting Mexico's central area to markets in the United States and Europe. Similarly, it will link the national sector to global supply chains, allowing it to become more competitive and meet the problems of global commerce and production (Market, Real Estate, 2017).

Export Movements by System, Mexico, 2013

Fig. 5.1 clearly shows that land transportation was the most popular mode of transportation in Mexico in 2013, followed by the sea. Sinaloa's top three exports are tomatoes, maize and wheat.

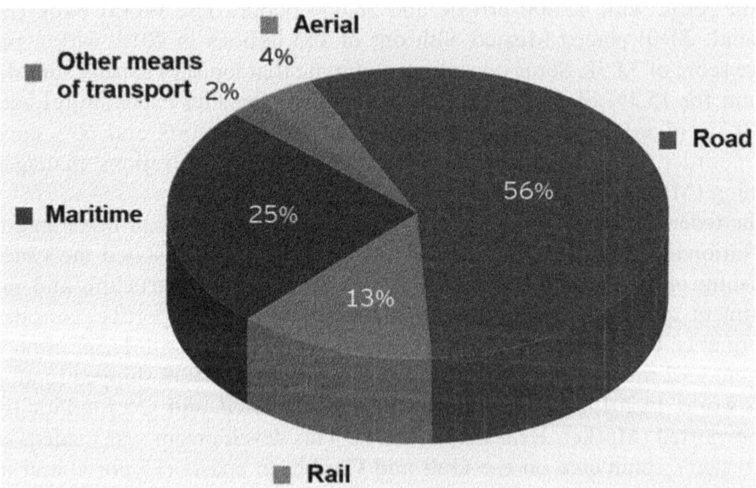

Fig. 5.1. Export Movements by System, Mexico, 2013. *Source:* Market, Real Estate (2017).

Of all Sinaloan exports to foreign nations, 82.5% are sales to the US market. This income accounts for 8.9% of the state's GDP. Sinaloa exported items worth around 770 million dollars in 2014, accounting for 0.2% of the national total. The principal exports were tomatoes, chickpeas, beef, fish oil and flour, prawns, squid and tuna. The NAFTA, which the United States plans to renegotiate, has resulted in a 339% increase in the value of Mexican exports to the United States since its introduction.

More than 80% of exports go to the United States. This indicates that the country's producing sectors are heavily reliant on its northern neighbour. And the more a state trades with the United States, the more vulnerable it is. Vegetables were the primary product from January to June 2010. Corn, soy, soy residues, wheat, pork ham, other goods, boneless meat, powdered milk and cream, fructose and starch waste were the top 10 North American imports into our nation over the same time period.

Mexico has 22 states that export veggies to the United States. Guanajuato was the leading state in terms of exports, with $583,795 million. Vegetables were imported from the United States by 21 other states for a total of $23 million, 139 thousand. In 2014, Sinaloa sent veggies to its northern neighbour worth $12 million, 139 thousand, 433 dollars. It then imported American veggies for $ 84,628. Thus, the value of vegetables exported by 22 states to the United States surpasses their imports.

The bilateral relationship is crucial. If the US market is crucial to Mexico, the same is true in reverse. According to the United States Department of Commerce, Mexico is the second-largest destination for exports from the country, accounting for 15.7% of total exports, followed by Canada (18.6%), and the third-largest provider, accounting for 13.1% of imports, after China (21.5%) and Canada (13.2%). For example, Mexico is Michigan's second-largest commercial partner (the country receives 31.03 % of that North American entity's total exports). In the instance of California, the Mexican market imports 12.53% of the things that the organisation exports. For Nebraska, the figure is 14.53%.

Vegetable Exports and Imports, Mexico-EU

According to the National Bureau of Economic Research (NBER), US content accounts for over 40% of Mexican exports to the American market. Renegotiation, under a magnifying lens, the announced amendment of the FTA, captures the attention of our country's economic sectors. Maintaining and expanding the accords is an important issue for several governments that rely heavily on the US market for exports.

In the case of Sinaloa, this commercial tie accounts for 8.9% of the state's GDP and is mostly related to food. There are other, more susceptible entities. Examples include Chihuahua, Baja California, Coahuila and Tamaulipas. Exports to the United States account for more than half of all these states' GDP and are mostly from the manufacturing industry (El Debate, 2017).

Research Limitations

During the preparation of this research, there were time constraints because it was carried out over a 7-week period and the research period could not be exceeded, a financial constraint was discovered because the necessary financial resources were not available to conduct field research, data collection because the variable 'Y' which is agricultural exports created it difficult to gather information and accuracy due to the data that were collected.

Results

As a summary of the results of the study, it was found that Mexico had quarterly GDP growth at annual rate in the past 3 years with a steady 2%, GDP per capita has been increasing until 2014, when it has been decreasing, it has a behaviour of the primary sector that has been in constant change for 2015, down to minus one, but for 2016, it increased up to 6% in accordance to the financier, Mexico's exports to the US have grown by 339% in terms of real since the implementation.

Mexico–US trade has been in constant flux; it is currently at 81%. In 2014, the Mexican states supplied to the United States for a total value of 583 million 794 thousand 468 dollars, Sinaloa for 12, 139, 4,443 dollars, with the three exported vegetables being tomato, corn and wheat. Mexico received a smaller quantity of US vegetables for a total value of 23 million 139 thousand 098 dollars and Sinaloa exports to the US 82.5%.

Graneros United leads the Sinaloan exporting enterprises with 34, 893, 112.75 (USD) and a participation of 1.35 and is ranked 12th overall. According to Council for the economic development of Sinaloa (Consejo para el desarrollo Económico de Sinaloa, CODESIN), Agroindustrias de Andar de Delicias SA de CV is in second place with 28, 883, 344.53 and a participation of 1.13 in the general position 14, followed by Agroexportadora del Noroeste SA de CV in third place with 24, 127, 318.57 and a participation of 0.94.

According to the World Bank, Mexico ranks 55th out of 155 nations in transportation performance and 15th in port technical efficiency in 2010. According to banker Yuridia Torres, Mexico has the potential to develop and extend its market to other nations such as China, Australia, Peru, Chile, Guatemala, Portugal, Turkey, Russia, Indonesia, Iran, Singapore, Vietnam, Taiwan, Japan and the European Union. It would be more advantageous if the TMEC renegotiation did not meet Mexico's expectations.

This research questioned if the renegotiation of the TMEC had a detrimental influence on Sinaloan agricultural exports; nonetheless, after conducting an in-depth examination, it was decided that if the United States decides to depart the treaty, it will have a significant impact on Sinaloa. United States as the country's main export destination, with 82.5% of sales to its market, but this is no longer a fact with a higher probability because President Donald Trump is looking at renegotiation after the new elections in Mexico in 2018 to further simplify the procedures that are necessary and not the final solution, leading us to conclude that the hypothesis is negative.

Renegotiation, along with the benefits and drawbacks, has a direct link with agricultural output in Sinaloa, with a significant negative impact, since overproduction would result if the new location for exporting from Sinaloa was not promptly accessible. Renegotiating, along with the advantages and disadvantages, has a direct relationship with agricultural trade in Sinaloa; it would negatively affect because Sinaloa would be left without its main export destination, but it would have the option of opening up to send its products to new locations like China, Peru, Chile, Guatemala, Portugal, Turkey, Russia, Indonesia, Iran, Singapore, Vietnam, Taiwan, Japan and the European Union, which are countries that in 2016 open.

The renegotiation, along with the benefits and drawbacks, has a direct impact on Sinaloa's agricultural distribution network; it would also have a significant impact on sea and land ports, which are the most common transportation routes in Mexico, with auto transport being used more frequently and seaports, the most popular being in Veracruz.

Discussion

Mexico is a highly capable country, with sufficient assets to stand out from other countries in Latin America and the world; it is only a matter of having an efficient president who understands how to take advantage of and exploit the country's riches, who focuses on persevering and growing every day to become a developed country, improving the quality of life of its citizens. Mexico should not be so reliant on the United States, since it can sustain its economy even without it. It must overcome biases and anxieties and export its products to new markets, taking risks outside of its comfort zone to achieve even better outcomes for the Mexican economy.

In Sinaloa, while the activities belonging to the industrial and services sectors registered a decrease of 6.5% and 6%, respectively, the activities corresponding to the primary sector had a growth of 8.1% in 2020 (Codesin, 2021). In this sense, Sinaloa producers continue to take advantage of the business opportunities offered by the national and international agri-food markets, maximising the efficiency of production processes, the responsible use of natural resources, social responsibility, modernisation and expansion of capacity installed of its productive assets and the use of financial services that complement its investment needs (Reyes Rodríguez & Quintero Reyes, 2020).

The main component analysis establishes that in the state of Sinaloa, the most relevant crops in terms of planted area are cereals, fruit trees, industrial plants, legumes and tubers. The harvested area is generally similar to that sown, except that industrial crops disappear from the first component. Regarding the volume of production, cereals, fruit trees, forages, tubers and oilseeds are the most important in Sinaloa (Bruno et al., 2017).

Thanks to the fact that Sinaloa in its economy has strength in the agri-food, agriculture, agroindustry and mining sectors, it managed to cushion the fall in the export indicator of its different products for the month of April. Even though at the national level, the drop in total exports was −40.9%, Sinaloa registered −14.4% in April thanks to the agri-food, agriculture, agribusiness and mining sectors, so we will continue to promote our exports and manage in favour of the Sinaloan producers (Redacción, 2020).

The analysis of the evolution of the agricultural production structure of Sinaloa since NAFTA and later TEMEC (USMEC) shows a growth in the production of fruit trees and oilseeds, that is, crops with an export vocation, to the detriment of peasant economy crops. The growth of the basic products of the state food basket, although it has been constant, is not enough to cover all the needs of the internal market, which generates a crisis to satisfy a part of the national demand, which was covered with surplus production (Bruno et al., 2017).

The year 2020 was not only full of challenges, but also many opportunities. The approval of the United States of North America with the FTA gives certainty to Mexico and with it great positive repercussions to the various sectors and where Sinaloa reaches important respite. But on the other hand, the pandemic also had negative effects on product exports (Toledo, 2020).

This research can be extremely beneficial to Sinaloa's main agricultural exporting companies, as well as various secretaries such as the secretary of economy, SAGARPA, Secretary of Economic Developement (Secretaría de Desarrollo Económico, SEDECO), National Service of Health, Safety, and Agri-Food Quality (Servicio Nacional de Sanidad, Inocuidad, y Calidad Agroalimentaria, SENASICA), For México (Pro Mexico), Chambers of Commerce and students pursuing degrees in Economy and International Commerce at the Autonomous University of Sinaloa (UAS).

In this perspective, the extension of the Veracruz port complex, which is scheduled to be finished in 2025, represents public policy aimed at reinforcing the port network. With a total investment of 60,000 million pesos, including 23,933 million pesos between 2013 and 2018, the port will reach the maximum number of marine entries and freight mobility under the National Infrastructure Programme 2014–2018. This will result in an extension of the northern port in the following areas: breakwater, navigation channel, piers and specialised terminals, specialised equipment, new yards and storage area.

There were several constraints in this research owing to the difficulty of acquiring particular information while gathering the ideas that defined the variable 'Y', which is known as agricultural exports and the limited amount of time available to complete the research project.

Conclusion

As a general conclusion of this research project, it is argued that the renegotiation of the TMEC does not have to cause a wave of uncertainty regarding agricultural commodities and products from the State of Sinaloa's exports because the renegotiation was accepted and a fair agreement was reached for the three countries. This agreement includes many terms that safeguard agricultural exports from any planned unfair behaviour. Furthermore, this prompted us to look into new opportunities for agricultural export development in Mexico, since countries such as China, Australia, Singapore and others have opened their markets to agricultural commodities imported from Mexico. Finally, the results are favourable for the further expansion of agricultural goods from Sinaloa to Canada and the United States, the TMEC partners.

References

Audley, J. J., Demetrios, G., Papademetriou, D. G., Polaski, S., & Vaughan, S. (2003). *La promesa y la realidad del TMEC, lecciones de México para el hemisferio*. Recuperado el 23 de 07 de 2017. http://carnegieendowment.org/pdf/files/NAFTA_Spanish_fulltext.pdf

Banco Mundial. (2010). *PIB y PIB per cápita de México*. Banco Mundial. Consultado el 05/07/20177en. https://datos.bancomundial.org/indicador/NY.GDP.PCAP.CD

BANXICO. (2017). *Desarrollo económico mexicano*. Banco de México. Recuperado el 23 de 07 de 2017. http://www.banxico.org.mx/politica-monetaria-e-inflacion/estadisticas/graficas-de-coyuntura/produccion-ventas-y-precios/producto-interno-bruto–varia.html

Bruno, F. C., Restrepo, B. L. F., & Holmes Rodríguez, E. (2017). Estructura productiva agrícola del estado de Sinaloa, México, y el tratado de libre comercio de américa del norte (tlcan). *Chilean Journal of Agricultural & Animal Sciences*, *33*(1). Chillán mayo 2017. http://dx.doi.org/10.4067/S0719-38902017005000102. https://scielo.conicyt.cl/scielo.php?script=sci_arttext&pid=S0719-38902017000100102

Castro, A. (31 de 01 de 2008). Que es el tratado de libre comercio. *El Universal*. Recuperado el 15 de 06 de 2017. http://archivo.eluniversal.com.mx/notas/478522.html

CEDRSSA. (2020, Julio). *Reporte del sector agropecuario en México. Centro de Estudios para el Desarrollo Rural Sustentable la Soberanía Alimentaria*. Poder legislativo de San Lázaro, Cd. De México. http://www.cedrssa.gob.mx/files/b/13/22Situacion_Sector_Agropecuario_Me%CC%81xico.pdf

CODESIN. (2017). *Las exportaciones en Sinaloa*. Consejo para el Desarrollo Económico de Sinaloa. Recuperado el 21 de 07 de 2017. http://codesin.mx/news/las-exportaciones-en-sinaloa/:%20codesin.mx

Codesín. (2021). *Reporte 07 del 2021. Indicador trimestral de la actividad económica estatal (ITAEE) en Sinaloa*. Consejo para el Desarrollo Económico de Sinaloa. https://sinaloaennumeros.codesin.mx/wp-content/uploads/2021/01/Reporte-07-del-2021-ITAEE-tercer-trimestre-2020.pdf

CONAPO. (2017). *Estadisticas de exportación* (G. D. Mexicana, Ed.). Consejo Nacional de población. Recuperado el 23 de 07 de 2017. http://www.conapo.gob.mx/en/CONAPO/Mexico_en_cifras

Dillon, J. (01 de 01 de 1991). *Bases teóricas y prácticas de los tratados de libre comercio ulc/TMEC/gatt/omc* (E. D. Castro, Ed.). Economista da Institution Ecumencial Coalition for Economic Toronto. Recuperado el 20 de 06 de 2017. http://www.users/Usuario/Downloads

El Debate. (05 de 03 de 2017). Exportaciones de Sinaloa. Mazatlán, Sinaloa. *El debate*. Recuperado el 15/07/2017. https://www.debate.com.mx/

El Financiero. (21 de Julio de 2017). Economista. *el financiero*. Recuperado el 23 de 07 de 2017. http://www.elfinanciero.com.mx/economia/el-pib-de-mexico-en-cinco-graficas.html

Galeano, E. (1970). *Las venas abiertas de América latina* (la cueva ed., Vol. II) (R. Villagra, Ed.). Siglo XXI editores. Recuperado el 20 de 07 de 2017.

Market, Real Estate. (2017). Artículos, Infraestructura y construcción, puerta para la industria-mexicana. *Market, Real estate*. Recuperado el 21 de 07 de 2017. http://www.realestatemarket.com.mx/articulos/infraestructura-yconstrucción/15387-puerta-para-la-industria-mexicana. http://www.realestatemarket.com.mx/

Ministro de Asuntos Exteriores y de Cooperación. (junio de 2017). *Exteriores de gobierno*. Ministro de asuntos exteriores y de cooperación. Recuperado 20/07/2017. http://www.exteriores.gob.es/representacionespermanentes/espanaue/es/quees2/Paginas/El-Derecho-comunitario.aspx

Mypimes. (2017). *Libre comercio agrícola, teoría y realidad*. Mypimes. Recuperado el 25 de 07 de 2017. http://www.tlc.gov.co/publicaciones/12942/libre_comercio_agricola_teoria_y_realidad. http://www.tlc.gov.com/

NAFTA de hoy. (01 de 01 de 2014). Tratado de libre comercio de américa del norte. *TMEC de hoy*. Recuperado el 19 de 07 de 2017. https://www.sec-TMEC-mex.org/Inicio/Bienvenida. http://www.sec-tmec-mex.org/

Redacción. (9 de junio de 2020). Gobierno de Sinaloa fortalece sectores primarios. *El Economista*. https://www.eleconomista.com.mx/estados/Gobierno-de-Sinaloa-fortalece-sectores-primarios-20200609-0138.html

Reyes Rodríguez, E., & Quintero Reyes, G. (2 de abril de 2020). Agronegocios Oportunidades de inversión en el sector agroalimentario en Sinaloa. *El Economista*. 02 de abril de 2020, 19. https://www.eleconomista.com.mx/opinion/Oportunidades-de-inversion-en-el-sector-agroalimentario-en-Sinaloa-20200402-0111.html

Secretariado del TLCAN. (22 de 07 de 2017). NAFTA. *Secretariado del TMEC*. Recuperado el 14/06/2017. https://www.nafta-sec-alena.org/Inicio/Introducci%C3%b3n

Toledo, M. (2020). Sinaloa, en el primer lugar en exportación de productos cárnicos. *El Sol De Mazatlán*. Lunes 27 de enero de 2020. https://www.elsoldemazatlan.com.mx/local/sinaloa-en-el-primer-lugar-en-exportacion-de-productos-carnicos-4756437.html

Torres, Y. (20 de 02 de 2017). 14-paises-abrieron-su-mercado-a-productos-mexicanos-en-2016. *El financiero*, Recuperado el 19 de 07 de 2017. http://www.elfinanciero.com.mx/economia/14-paises-abrieron-su-mercado-a-productos-mexicanos-en-2016.html. http://www.elfinanciero.com.mx/

Chapter 6

Role of Crop Insurance in Sustainable Agriculture Practices: A Global Perspective

Saheli Bhattacharya and Bappaditya Biswas

University of Calcutta, India

Abstract

Crop insurance protects the farmers and helps in reducing any element of chance associated with agriculture. It is a contract between the agriculturist and the insurer where the insurer promises to pay the farmer an indemnity in case of loss suffered by him due to unforeseen events, against a premium. Crop insurance creates a sense of confidence among the farmers which in turn encourages them to follow the agricultural best practices, thus taking a step towards sustainable agriculture. The concept of sustainable agriculture can be explained as producing food and livestock by causing minimum damage to the environment. Crop insurance thus creates a win-win situation for sustainable agriculture. The importance of crop insurance is undeniable and that is the reason its global footprint is quite prominent. The size of crop insurance market was USD 39.81 billion in 2022, and it was expected to grow to USD 61.51 billion by 2030. Among all the regions, Asia–Pacific is expected to grow at a very fast pace. The objectives of the present chapter are to explain the role of crop insurance towards sustainable agriculture and to make a study of the crop insurance schemes prevalent in selected European and Asian countries. The data have been collected from various secondary sources like books, journals, websites, etc. The authors have come to the conclusion that crop insurance is being offered in the selected European and Asian countries in various forms. Public–private partnership is very common among these selected countries.

Keywords: Crop insurance; insurance products sustainable agriculture; public–private partnership; European countries; Asian countries

Growth, Poverty and Developmental Aspects of Agriculture, 79–92

doi:10.1108/978-1-83608-076-320241006

Introduction

Crop insurance, simply put, is a contract between the insurer and the insured which is taken up by the farmers (i.e. insured) for protecting their agricultural produce from various perils they might be exposed to. The concept of crop insurance was first introduced in European countries like France and Germany in the 18th century. The importance of crop insurance to the farmers is undeniable. Crop insurance assures income for the farmers, encourages farmers in choosing risky crops for production, provides insulation form financial distress and also encourages farmers in adopting better production techniques.

Crop insurance is offered in different forms across the globe. Though, broadly speaking, there can be three systems of crop insurance namely traditional indemnity-based system, index-based system and hybrid system. Within the purview of traditional indemnity-based system comes named peril crop insurance (for example, hail insurance) and multi-peril crop insurance (MPCI). As the name suggests, index-based crop insurance schemes refer to the schemes where the indemnity payment is based on some index. Some examples of the index-based crop insurance schemes are weather index, yield index, normalized difference vegetation index (NDVI), etc. Hybrid system of crop insurance, on the other hand, is a combination of both named peril crop insurance and weather-based indices.

Across the globe crop insurance is offered either compulsorily or voluntarily to the farmers. Most of the countries offer crop insurance voluntarily though few countries have made the participation of the farmers in the crop insurance schemes compulsory. The models of crop insurance followed internationally are either private model of crop insurance, public–private partnership (PPP) model or the public model of crop insurance.

Crop insurance market is a growing market, the fact that its market size was USD 39.81 billion in 2022, and it is expected to grow to around USD 61.51 billion by 2030, only establishes the same (https://www.zionmarketresearch.com/report/crop-insurance-market#:~:text=In%20terms%20of%20revenue%2C%20the,to%20changes%20in%20climate%20conditions Retrieved on 19.02.24). Since Asian countries boast large numbers of agricultural land, the highest growth of global crop insurance market is expected to happen in the Asia–Pacific region. In recent times, a sudden increase in the demand for crop insurance has been observed globally; the reason behind the same might be sudden and severe climatic changes. Crop insurance along with other sustainable agricultural practices may help in taking a step towards agricultural sustainability.

Literature Review

Ahsan et al. (1982) explained the various aspects that were to be taken into consideration for formulation of a public insurance scheme. According to Ifft (2001), the government-managed crop insurance schemes were unsuccessful globally. So he felt that the private players should be given more opportunity. Jain (2004) was of the view that private sector would not be much interested in offering agricultural insurance products as there was not much scope of earning profits, yet they should be encouraged to the extent they were willing to offer. Gangopadhyay (2004) observed that crop insurance schemes should be offered and managed by the non-government organizations (NGOs).

Sherrick et al. (2004) noted in his study that the respondents though were concerned with risk management techniques, but they did not consider crop insurance as an important one. Niewuwoudt and Bullock (1985) observed in spite of the subsidies provided the demand for crop insurance were quite low. According to them the demand for crop insurance was dependent on factors like farm size, expected risks, expected rate of return on insurance, etc. Zacharias (2011) was of the view that there would be massive change in the world by the year 2020, and he felt that it would impact the demand for crop insurance schemes.

Prabhu and Ramackandran (1986) pointed out various important dimensions of crop insurance schemes prevalent in different countries. He found that demand pattern of crop insurance may vary region to region and crop to crop, so a single scheme is not sufficient, and since crop insurance is linked to credit, the poor access of small and marginal farmers to institutional credit system may pose a hindrance in fulfilment of the primary objective of crop insurance.

Nair (2010) has rightly pointed out that the cases of agricultural suicides across the country imply that risk mitigation schemes have major shortcomings; though farmer suicides can be attributed to various reasons, but a successful insurance programme could reduce the pain of the farmers to a great extent in the crisis years. In the article, she debates which of the traditional or index-based insurance is better and is of the view that both has its advantages and disadvantages, so a blend of both may be the need of the hour.

Höppe (2007) in his article shows that global warming is increasing the risk of weather-related catastrophes like windstorms, floods, droughts, etc. And if no measures are adapted to combat the changing weather patterns, then agriculture will surely be affected. Proper insurance systems can help farmers to cope with the increasing volatility of their losses.

Rao (2010) in his book made an extensive study regarding the crop insurance mechanism prevalent in countries like Brazil, United States of America, Israel, Canada, China, Japan, etc. In *Brazil*, the private sector agricultural insurance is voluntary in nature. The public sector PROAGRO (Guarantee Programme for Agriculture and Livestock Activities, Programa de Garantia da Actividade Agropequaria) and SEAF (Insurance for Family Agriculture, Seguro da Agricultura Familiar) crop credit insurance programmes are compulsory for crop credit recipients of PRONAF (National Programme for the Strengthening of Family Agriculture, Programa Nacional de Fortalecimento da Agricultura Familiar). In *Canada*, various indemnity-based and index-based insurance programmes are offered. Indemnity based insurance products include crop hail, MPCI and livestock accident and mortality insurance among others. Among the index-based products offered are weather-index insurance, satellite imagery-based NDVI insurance and area-yield insurance. In *Israel*, KANAT (Insurance Fund for Natural Risks in Agriculture Ltd) provides cover for almost all crops, livestock and aquaculture. They cover key weather perils, including drought in arable crops and limited disease cover. Quantity and quality loss is covered, and livestock and poultry are covered against mortality. In *China*, though agricultural insurance schemes existed since 1982, there was a complete overhaul in the schemes in 2008. The market consists of a mix of general insurers, specialist agricultural insurers, mutual and private insurers and pools. The majority of China's crop insurance is individual grower MPCI. Though some products based on named-peril crop insurance are also available. In *Japan*, the Agricultural Insurance Scheme is based on the principle of solidarity among farmers. It relies on a network of cooperatives at the local, regional and national levels. The insurance products available under the Japanese agriculture insurance schemes are specified by law.

Salinger (2007) explained that the uncertainties associated with the crop yields, soil salinity, water erosion, etc. had increased in the Southwest Pacific region due to the changes in climate and extreme climatic events. Čolović and Petrović (2014) observed that there could not be one best suited model of crop insurance as it had to be decided based on the requirement of the state and the agricultural producers. Thapa and Adhikari (2018) were of the view that not only did crop insurance provide a safety net to the farmers but also encouraged them to adopt new farming technologies which in turn helped in enhancing the overall productivity.

Objectives

The objectives of the present study are:

- To discuss the role of crop insurance towards sustainable agriculture.
- To make a study of the crop insurance schemes prevalent in selected European and Asian countries.

Methodology

The data have been collected from various secondary sources like books, journals, websites, etc. The present study is divided into three parts. The first part of the chapter explains the role of crop insurance towards sustainable agriculture. The second part is divided into two sections. In the first section, the authors have discussed the crop insurance schemes prevalent in selected European countries, and in the second section, the authors have discussed in details the crop insurance schemes prevalent in the selected Asian countries. Lastly, in the third part, the authors have summarized the findings by presenting it in a nutshell and also drawn conclusions.

Role of Crop Insurance Towards Sustainable Agriculture

Sustainable agriculture or farming may be defined as production of food and livestock by causing minimum negative effect on the environment. In other words, providing food to the human population but not at the cost of harming the environmental ecosystem. Sustainable agricultural practices are need of the hour. Minimization of the effects of climate changes is possible by adopting eco-friendly farming methods. Sustainable farming practices can be achieved by the farmers in many ways. It could be done by cover cropping, relying on organic farming methods, reducing usage of synthetic chemicals, etc. Efficient utilization of resources is another pillar of sustainable agriculture.

Crop diversification can ensure sustainable agriculture. Crop diversification refers to a situation where the farmers cultivate a variety of crops thus reducing their risks against adverse weather conditions. Thus, crop diversification not only promotes healthy ecosystem but also contributes to its overall stability. Sustainable agriculture may also be achieved through maintaining soil health. Thus, using organic fertilizers and cover cropping becomes very crucial in maintaining the soil health.

Enhancing agricultural development's sustainability is also impacted by a greater use of renewable energy sources, such as biofuels. Long-term dependency on oil and gas corporations can be reduced, which will lessen reliance on external market circumstances and price swings for gasoline and diesel. In one study, different technologies for oilseed crop production are examined. These technologies allow for the optimization of crop selection as well as cultivation technology, taking into account the features of the land, the equipment available, and the crop's potential yield.

In light of the above, one might wonder about the role of crop insurance in sustainable agricultural practices. Sustainable farming practices may require investments in new and advanced technologies. Crop insurance acts as a safety net for the farmers and helps in shielding themselves against natural calamities, pest attacks, etc. thus providing them with the confidence in making such heavy investments. Agricultural stability is dwindled by the unpredictable weather conditions. Crop insurance protects the farmers against losses due to these climatic changes, hence managing the income fluctuations faced as a result of these changes.

Agricultural insurance lowers the quantity of chemicals discharged into the environment while simultaneously encouraging farmers to use more ecologically friendly agricultural practices and increasing production efficiency. Air pollution, the total production of ecologically friendly agricultural inputs and agricultural insurance have a significant correlation, according to data from 50 US states. The production of organic goods may rise when agricultural insurance is expanded.

The insurance sector plays a key role in supporting the transition to a more sustainable and climate-resilient agriculture. By developing new insurance products and investing in nature-based solutions (NBS) projects, the insurance industry can help reduce the risk of extreme climate events to farmers and other agricultural stakeholders. Insurance industry participation in the development of NBS-based agricultural products is a multifaceted approach to compliance. These products promote sustainability, sustainability and economic stability by promoting cooperation and education in the agricultural sector. This collaboration between the insurance industry and agriculture is a promising way to create a more sustainable and sustainable future.

Below mentioned are some of the pointers which may help in developing a sustainable agricultural insurance system which in turn is essential for sustainable agriculture:

- Information technology will help the agriculture industry grow and remain sustainable, which will benefit insurers as well as farmers, lenders, traders, economists, climate scientists, legislators and agricultural researchers.
- It is very crucial to technically determine the risk exposure and risk appetite of the farmers as it is reflected in the premium and coverage.
- PPPs could be established, which helps in improving the efficiency of the agricultural insurance.
- The farmers need to be empowered to manage their own business risks.

For sustainable agriculture, crop insurance is essential as it promotes holistic risk management mechanism. The financial security promised by crop insurance in the event of crop failure makes the farmers maintain economic stability and also helps in recovery of the investments. Convergence of sustainable agricultural practices with crop insurance is required a much larger scale now. Governments and insurance providers could consider reducing insurance premium for the farmers engaged in sustainable agriculture.

Crop Insurance Scenario in Selected European and Asian Countries

Crop insurance is offered across the globe. The figure (Fig. 6.1) shows the agricultural insurance premium values of different regions worldwide for the years 2007 and 2017:

As it can be seen from Fig. 6.1, there has been a drastic shift in the region-wise percentages of agricultural insurance premiums between the years 2007 and 2017. 63.50% of the global insurance premium was written in North America in 2007; the percentage dropped to 40% in 2017. In 2007, the second spot was held by Europe (16.90%) which was closely followed by Asia–Pacific (15.70%). Whereas in 2017, Asia–Pacific held the first position at 42%, and the percentage of agricultural insurance premium written in European countries was 13%.

European Countries

European countries like France and Germany were the flag bearers of crop insurance schemes. The first crop insurance schemes which were introduced in these countries date back to 1820s when hail insurance was introduced to safeguard the grape growers. Later on, crop insurance schemes were adopted in most of the countries across the globe. The following paragraphs give a brief account of the crop insurance scenario in selected European countries:

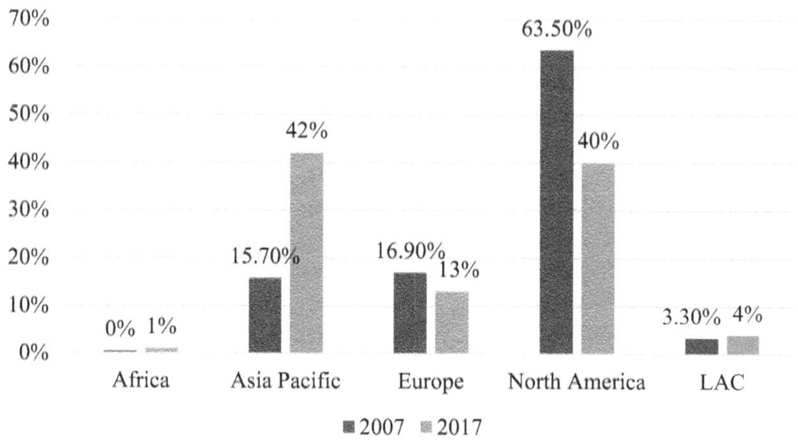

Fig. 6.1. Region-Wise Agricultural Insurance Premium for 2007 and 2017. *Source:* https://www.giz.de/expertise/downloads/2021%20GIZ_Innovations%20and%20emerging%20Trends%20in%20Agricultural%20Insurance-An%20update.pdf Retrieved on 08.02.24.

France

France was one of the first countries to adopt crop insurance. Though crop insurance was first offered in 1900s in France, but until 1964, there was no state sponsored agricultural insurance programme. The compensation of loss was made by the Government of France on case-to-case basis. In 1964, the National Agricultural Disaster Fund (FNGCA) was set up by promulgation of the Law 706/1964. The primary aim for creation of the fund was hedging the farmers against any natural disasters. Since the fund was jointly financed by the government and the farmers, they shared the premiums equally.

The agricultural insurance system and market in France has witnessed many changes over the years. Private insurance market also gained momentum since 2005 as the French government decided to accelerate the private insurance.

Series of natural disasters in the recent years led to the announcement of a universal crop insurance system in the year 2021 which was to be implemented from the year 2023. There would be three levels of compensation; for the lowest risks, there would be self-insurance; for medium risks, reimbursement would be done by private insurers, and the serious risks would be covered by the public funds. Within 5 years, the aim is to increase the number of insured producers to 60% from 30%

Germany

Though Germany has a long history of crop insurance (hail insurance was first introduced in 1733 and livestock insurance in 1830), yet till 2006, there was no developed agricultural insurance in the country. The most important product offered in the country is still hail insurance; around 60% of the total crop area is covered under it and covers all arable crops including vegetables, vineyards and fruit plantations. The MPCI was introduced only recently, still in nascent stage and is quite underdeveloped possibly due to strict underwriting conditions prevalent in the country. Only 0.5% arable land was insured as of 2015 against accumulation risks.

Index-based drought insurance products have been made available for the farmers only recently and are offered for winter wheat, winter oilseed canola, winter rye, beets, potatoes and silage corn. Greenhouse insurance is also offered in the country. In Germany, forestry insurance and aquaculture insurance are also offered. Livestock insurance is offered, and coverage is available for accident, fire, animal disease like foot and mouth disease and cow disease, epidemic diseases and movement restrictions. Livestock insurance is of extreme importance in the country and is widely accepted by the farmers. More than 50% of the farmers are covered under it.

The crop insurance market in Germany is extremely competitive as both private and public mutual insurance companies actively participate. Crop hail insurance is offered by 14 companies, and MPCI is offered by only one company. The crop insurance products are delivered to the farmers through cooperatives and farmer's associations. No government support is available for crop insurance in the country. The premiums are not subsidized by the Federal government of Germany probably because it would require huge annual funding. No form of public support is available for reinsurance.

Spain

Spain too has an illustrious history of crop insurance. There were a few companies operating in Spain in the beginning of 20th century who insured loss of harvest due to the risk of fire. Different systems of agricultural insurance were offered with varying degree of participation from the state between 1920s and 1970s. Since 1978, Combined Agricultural Insurance System is being offered in Spain. It is based on mixed system, i.e. PPP.

The public sector, represented by Entidad Estatal de Seguros Agrarios O.A. (ENESA, Ministry of Agriculture, Fisheries and Food), and private sector, represented by Professional Agrarian Organizations Agri-food Cooperatives, private reinsurers and the insurance sector through AGROSEGURO (Agrupación Española de Entidades Aseguradoras de los Seguros Agrarios Combinados), actively participate in the crop insurance market. AGROSEGURO is a pool of private insurance companies which participate in the co-insurance; presently, 29 insurance companies (including Consorcio de Compensacion de Seguros) participate in the pool. The participating companies assume the risk in the same proportion as of their share capital in AGROSEGURO. In 2023, after the recent drought crisis in the country, AGROSEGURO is bracing for a total pay-out of around € 300 million (US$ 325 million).

Crop insurance is offered in three models. The first model is farm insurance (Module 1 covered serious damage resulting in 30% or more production loss, Module 2 main risks for the products and exceptional damage causing losses in each plot and remaining risks causing damage in the entire holding and Module 3 all risks causing losses in each plot were compensated). In the second model, there is a basic insurance policy, and along with it, there is an additional guarantee covering the catastrophic damages. The third model is combined, where both farm insurance and basic insurance coexist though it excludes the 'other adverse weather phenomena'.

Asian Countries

As already highlighted before, the need of crop insurance is growing at a fast pace in the Asian countries due to the large number of arable land. Region wise, Asia–Pacific was the leading revenue generator in 2022. Globally, China and India are among the top three countries which have a comprehensive and extensive crop insurance ecosystem.

China

In China, approximately 40 million ha crops are destroyed every year as a result of hail, typhoon, diseases, insect pests, flood, drought and other perils. The total loss due to these perils amounts to almost US$ 12.5 billion (CNY 100 billion) annually. China has an illustrious history of crop insurance. The country's agricultural insurance history can be divided into two broad phases.

The first phase began in 1982 and continued till 2002; in this phase, both crop and livestock insurance were introduced. In this phase, People's Insurance Company of China (PICC) developed agricultural insurance products and delivered them to farmers in rural areas through local government. The results of agricultural insurance were very unsatisfactory during this period as the insurance products proved to be highly unprofitable, and thus, the insurers lost interest in selling them further.

The second phase began in 2003; various new and subsidized agricultural insurance programmes were introduced in this phase. The developments in this second phase were of extreme importance as it has actually led to the present market standing. PICC joined hands with local governments in launching several pilot agricultural insurance products in the market. Insurance coverage was provided by China United Property Insurance on pilot basis for rice, wheat and barley in Xinjiang province. Since the beginning of the second phase in 2003, four insurance companies (Sunlight, Anxin, Anhua and Groupama) have started offering agricultural insurance products besides PICC and China United Property Insurance.

In 2016, 26 insurance companies were offering agricultural insurance products in China. Of these 26 companies, six companies (namely, PICC, China United Property Insurance Company, Sunlight Agricultural Mutual Insurance Company, Guoyuan Agricultural Insurance Company, Anhua Agricultural Insurance Company and AVIC Groupama Insurance) dominated and held around 93% of the agricultural insurance market share. Again, among these six companies, PICC was the major agricultural insurance provider and held around 54% of the agricultural insurance market share.

In China, a wide variety of agricultural insurance products are available. Livestock insurance was available in the country in 2007, and it was offered by five of the leading insurance companies, namely PICC, CUPIC, Anxin, Anhua and Sunlight. The coverage of livestock insurance was available for reproductive sows against epidemic diseases (including blue ear, which is also known as porcine reproductive and respiratory disease) and other natural hazards like flood, fire and typhoon. Livestock price index insurance was introduced in May 2013; however, it was available only for pig farmers. Forestry insurance, greenhouse insurance and aquaculture insurance are also available in the country.

The government plays an active role in extending its support for agriculture and agricultural insurance. The government supports the agricultural insurance through premium subsidies, government technical agencies (which helps in loss assessment) and also acting as the last resort for reinsurance. When both the crop and livestock insurance were at pilot stages, the premium subsidies, depending on the province, ranged from 20% to 100%. However, in 2010 the maximum subsidy level was 65%. The government also provides support in the event of natural disasters, but the compensations are made on a case-to-case basis.

India

The concept of crop insurance was first mentioned in India in 1920. J. S. Chakravarthi of Mysore state first proposed a rain insurance scheme for protecting the drought inflicted farmers. But unfortunately, the scheme was not put into practice at that time. Though the question of formulating and implementing a crop insurance scheme was on the anvil since post-independence (i.e. 1947), but the first ever crop insurance scheme was introduced in the year 1972–1973.

The first ever crop insurance scheme which was introduced in 1972–1973 was discontinued in 1978–1979. Thereafter, many schemes, Pilot Crop Insurance Scheme (PCIS), Comprehensive Crop Insurance Scheme (CCIS), Experimental Crop Insurance Scheme (ECIS), National Agricultural Insurance Scheme (NAIS), Pilot Scheme on Seed Crop Insurance (PSSCI), Farm Income Insurance Scheme (FIIS), Weather Based Crop Insurance Scheme (WBCIS), Modified National Agricultural Insurance Scheme (MNAIS), Coconut Palm Insurance Scheme (CPIS), National Crop Insurance Program (NCIP), have been implemented in the country. Most recently, Pradhan Mantri Fasal Bima Yojana (PMFBY) has been introduced from 2016, and it is the ongoing crop insurance scheme of the country.

The programme encompassed oilseeds, annual horticultural/commercial crops and food crops (cereals, millets and pulses). Pilot coverage was available for perennial horticulture crops, for which a standard yield estimation methodology was available. The scheme is compulsory for the loanee farmers (i.e. the farmers who are taking loan from financial institutions) and voluntary for the non-loanee farmers. The scheme was available for all the sharecroppers and tenant farmers who were growing the notified crops in the notified areas.

The rate of premium was to be determined using an actuarial basis, and the Central Government and the State/Union Territory (UT) Government would split the difference 50:50 between the actuarial rate and the rate of premium that the farmer would be required to pay. The State/UT governments were granted the freedom to increase the amount of subsidy over the standard subsidy from their budget; however, premium subsidies may only be increased up to the amount of the Sum Insured.

In the year 2018, the area insured (in '000 hectares) under PMFBY scheme was 47,625.21 hectares. However, the area insured (in '000 hectares) decreased over the years; for 2019, it was 44,684.39; for 2020, it was 42,911.58; for 2021, it was 38,736.39. The area insured (in '000 hectares) slightly increased in 2022 – 39,816.63, and there was a massive increase in 2023 – 47,057.19. The total claim paid (in lakhs) for the years 2018 and 2019 were Rs. 2,287,665 and Rs. 2,548,687. However, the amount of total claim paid (in lakhs) reduced over the next few years; for 2020, it was Rs. 1,931,603; for 2021, it was Rs. 1,805,935; and for 2022, it was Rs. 1,524,830. For 2023, the total claim paid (in lakhs) till now was Rs. 275,134 (https://pmfby.gov.in/adminStatistics/dashboard Retrieved on 23.02.24).

Indonesia

The 2013 national census data published by Badan Pusat Statistics (Central Bureau of Statistic) reveal that out of the total 26.14 million agricultural households of Indonesia, almost around 70% of the farmer households are engaged in rice production. The national census reported that around 39.96% of the rice producing farmer households had to bear losses due to pests and diseases in 2013. Not only due to pest attacks and diseases, agricultural households in Indonesia have to experience crop losses due to natural disasters like earthquakes, tsunamis, floods and droughts, given the country's location on the Pacific Ring of Fire. Climatic changes which increase the frequency of El Nino events also adversely affect Indonesian agriculture.

In 2009–2010, two pilot agricultural insurance programmes were introduced by Government of Indonesia through Ministry of Agriculture. The schemes were implemented in West and Central Java; one scheme offered MPCI and the other livestock mortality and theft cover. The scheme was linked to credit, and the delivery channels were through commercial banks and rural banks. The government had financed 100% of the premiums. But the government is thinking over three different models for agricultural insurance for implementation in the future. The first model has the premium subsidies fully subsidized by the government, the second model aims to create a commercial insurance environment which will link the insurance companies with the input suppliers and agribusiness and, lastly, the third model is a crop-credit linked insurance in which the farmers who want to take credit would have to compulsorily purchase insurance.

A flood weather index has been implemented in Indonesia since 2009, but the scheme is not for agriculture; it is for small and marginal urban property owners in Jakarta. Individual level rainfall deficit insurance has also gained importance in Indonesia since 2009 as the International Finance Corporation of the World Bank in association with the Australian Agency for International Development (AUSAID) has financed a feasibility study for the same. The study is for the maize farmers of the Negusa Tengara Barat (NTB) and East Lombok of East Java.

Since rice is an important crop for Indonesia and rice farming is the main source of livelihood for many small scale subsistence farmers, a pilot Asuransi Usaha Tani Padi (AUTP) insurance programme was developed by the Ministry of Agriculture for the planting season of 2012–2013. Several studies and surveys were conducted to understand the farmer's perception of agricultural insurance before implementation of the pilot scheme. The main aim for implementing the programme was to promote agricultural development and food security.

It was initially planned to implement the programme in three provinces, namely, South Sumatra, East Java and West Java; however, due to some technical reasons, the project was carried on in East Java and South Sumatra only. Under the scheme, an insurance claim could be made by the farmers only if the crop damage was equal to or exceeded 75% of the total planted area. The amount to be paid was determined by multiplying the number of damaged acres by the sum insured. The scheme was subsidized by the government, where the government bore 80% of the premium cost, and mere 20% had to be borne by the farmers. The pilot scheme was very successful which paved way for further schemes.

With the help of the Japan International Cooperation Agency (JICA), state-owned fertilizer businesses have implemented the rice crop insurance during the first pilot project, which was for the planting seasons of 2012–2013 and 2013–2014. In addition to the MPCI, which is currently in its experimental phase, Indonesia also offers plantation and forestry fire protection. Both commercial banks and rural banks offer the farm insurance products. Although there isn't a specialized firm for agriculture insurance at the moment, Daspindo, a local insurance broker, participates in the pilot programmes for livestock and crops alongside 10 other insurers. The farm insurance programme is one in which the government is heavily involved. There is no information available on public disaster assistance programmes or reinsurance.

Empirical Findings

We can gain an overview of the many agricultural insurance programs that are available in a few chosen European and Asian nations from the discussion above. The

Table 6.1. Summary of Agricultural Insurance of Few Selected European and Asian Countries.

Country Name	Introduced in	Agricultural Insurance Products Offered	Government Support
France	1964	Data unavailable	Yes
Germany	1733	MPCI, named peril insurance, greenhouse insurance, forestry insurance, aquaculture insurance, livestock insurance	No
Spain	1917	Named peril insurance, farm insurance, catastrophic damage insurance, combined insurance, livestock insurance, forestry insurance, aquaculture insurance	Yes
China	1982	Named-peril insurance, index-based insurance, MPCI, livestock insurance, forestry insurance, greenhouse insurance, aquaculture insurance	Yes
India	1972–1973	MPCI, index-based insurance, livestock insurance	Yes
Indonesia	2009–2010	MPCI, livestock insurance, forestry insurance	Yes

Source: Compiled by the authors.

choice of these countries was made on the basis of the available data. The table (Table 6.1) summarizes the details regarding crop insurance in a nutshell:

Thus, from Table 6.1, it is noticed that all the countries (except Indonesia) have a long history of crop insurance. Crop insurance is offered in various forms in these countries, though MPCI is offered in almost all the countries. The government plays an extremely important role in implementation of the crop insurance schemes in any country. Here, except Germany, in all the countries, the crop insurance schemes are supported by the government.

Conclusion

From the above discussion, it is evident that crop insurance can be seen as a win-win situation in relation to sustainable agriculture. Crop insurance ensures safety net for the farmers. At the same time, crop insurance encourages farmers to make investments and to follow the farming best practices which in turn promotes sustainable agricultural practices. Thus, most of the countries around the world acknowledges the importance of crop insurance and offers the same. As already highlighted above, crop insurance in Asia–Pacific region is growing at a very fast pace, and among the Asia–Pacific countries, India and China have very well-developed agricultural insurance market. Among the three Asian countries discussed above, China has the highest number of farmer covered and area insured (195 million and 166.23 million hectares, respectively). Followed by India where the number of farmers covered were 47.32 million and area insured was 51.93 million hectares and Indonesia, it was 1 million and 0.806 million hectares, respectively.

In the above countries, various crop insurance schemes are available to the farmers. PPP is followed in most of these six selected countries. The governments of these countries play a huge role in making the crop insurance schemes popular among the farmers. The governments of most of these countries provide unstinted support to the farmers.

References

Ahsan, S. M., Ali, A. G. A., & Kurian, N. J. (1982). Toward a theory of agricultural insurance. *American Journal of Agricultural Economics*, *64*(3), 520–529.

Čolović, V., & Petrović, N. M. (2014). Crop insurance-risks and models of insurance. *Economics of Agriculture, Institute of Agricultural Economics*, *61*(3), 561–573.

Gangopadhyay, S. (2004). An alternative to crop insurance. *Economic and Political Weekly*, *39*(44), 4763–4764.

Höppe, P. (2007). Scientific and economic rationale for weather risk insurance for agriculture. In M. V. K. Sivakumar & R. P. Mohta (Eds.), *Managing weather and climate risks in agriculture* (pp. 367–375). Springer.

Ifft, J. (2001). *Government vs weather, the true story of crop insurance in India*. Centre for Civil Society. Research Internship Papers-2001.

Jain, R. C. A. (2004, January–June). Challenges in implementing agriculture insurance and re-insurance in developing countries. *The Journal*, 14–23.

Nair, R. (2010, February 6). Crop insurance in India: Changes and challenges. *Economic and Political Weekly*, *XLV*(6), 19–22.

Niewuwoudt, W. L., & Bullock, J. B. (1985). *The demand for crop insurance* (pp. 655–667). http://dx.doi.org/10.22004/ag.econ.183028

Prabhu, K. S., & Ramackandran, S. (1986, October 18). Crop-credit insurance: Some disturbing features. *Economic and Political Weekly*, *XXI*(42), 1866–1869.

Rao. (2010). *Agricultural insurance*. Insurance Institute of India.

Salinger, J. (2007). Challenges to coping strategies in Agrometeorology: The Southwest Pacific. In M. V. K. Sivakumar & R. P. Mohta (Eds.), *Managing weather and climate risks in agriculture* (pp. 99–112). Springer.

Sherrick, B. J., Barry, P. J., Ellinger, P. N., & Schnitkey, G. D. (2004). Factors influencing farmers' crop insurance decisions. *American Journal of Agricultural Economics*, *86*(1), 103–114.

Thapa, P., & Adhikari, A. (2018). Importance, scope and status of agriculture insurance in Nepal. *Journal of Agricultural Extension and Rural Development*, *4*(1), 365–371.

Zacharias, T. P. (2011). *Crop insurance in 2020*. NCIS Special Reports 160539, National Crop Insurance Services (NCIS).

Chapter 7

Interpreting Potato Price Volatility Through GARCH Estimation: A Novel Evidence

Madhabendra Sinha[a], *Partha Mukhopadhyay*[b] *and Durlav Sarkar*[c]

[a]Visva-Bharati University, India
[b]Durgapur Steel Plant, India
[c]Raiganj University, India

Abstract

Uncertainty in commodity pricing is a major cause of concern for farmers of developing countries like India. To observe the decision-making process of farmers of Burdwan district of West Bengal, India, the generalized autoregressive conditional heteroscedasticity (GARCH) model is employed to measure the extent of volatility in spot prices of potato and also to observe the existence of seasonal effects; Agmarknet database provided by Government of India over the period of 2003–2019 has been used. The market price of potato decreased during its season of production and peaked during the off-season period. The result implies that the volatility of potato forecasting is tending towards the standard error correction in the long run and from 2003 to 2019; the trend of potato price was influenced by multiple events. The result indicates that the farmers must focus on short-run structural events in the potato market.

Keywords: Price volatility; potato; GARCH model; Burdwan district; West Bengal; India; commodity price; rural market

Introduction

Like other economic variables, frequent variations or fluctuations in the price of a commodity around its long-run trend are also considered volatility of the price of

Growth, Poverty and Developmental Aspects of Agriculture, 93–106

doi:10.1108/978-1-83608-076-320241007

that product (Apergis & Rezitis, 2011). The volatility of prices is a common feature of almost all agricultural commodities (Degiannakis et al., 2022). This price volatility frequently affects the market's demand and supply conditions. The frequent change in the demand and supply conditions produces either excess demand or excess supply for agricultural products at a time in the market (Degiannakis et al., 2022). The existence of extra demand or excess supply further leads to fluctuations in the prices of agricultural products. In the case of India, the farm crisis is deep-rooted; to mention a few, the high cost of agricultural production vis-à-vis the low price of agricultural commodities and the massive mismatch between demand and supply of agricultural commodities might be the crucial factors in this regard.

The first shock comes from the supply side, as the supply of crops significantly varies through the market arrivals of the crops. The market arrivals depend on production costs, transportation and cold storage, as almost all crops are perishable. The speculative attitudes of the farmers and traders also play a vital role in the fluctuations in the supply of crops. The farmers of the crops earn a lucrative profit if the total of all costs is less than the revenue received by selling the crop, and the farmers bear the loss if the revenue received is less than the total of all fees. In the case of crops, it can be observed that farmers make decisions based on the status of their gains in the previous period. If they earn profit in the last period/year from the production of a crop, they are very likely to repeat the same crop and allocate more resources to produce that crop in the current year. This leads to a rise in supply. Irrespective of the fluctuations in demand, the frequent fluctuations in supply lead to the volatility of crop prices.

Potato (*Solanum tuberosum*) is an important agricultural crop, which is also known as 'Vegetable King'. It occupies a third place as the staple crop after rice and wheat. Potato is a rich source of vitamins, carbohydrates and starch (besides being a source of trace elements). The cultivation method of potatoes is also straightforward, and its price is usually affordable. For this reason, it has become a prime food for the people of developing nations to meet the required energy. Potatoes in developing countries are almost half of the global production. Despite their beneficial effects, farmers consider potatoes a semi-risky crop, as they are susceptible to diseases. In addition, the supply of potatoes varies significantly across seasons with consistent demand.

In the states of Uttar Pradesh, Bihar, West Bengal, Punjab and Karnataka, potato is cultivated during winter from October of any year to March of the following year as a *rabi* crop. It is grown as *a kharif* crop in the states of Himachal Pradesh, Uttaranchal, Jammu and Kashmir during the rainy season, i.e. from June to August. There has been a steep rise in the number of cold storage potatoes during the last 50 years due to the increase in consumption and production of potatoes and for storage during its harvesting seasons to meet its demand in the off season. The Burdwan district of West Bengal is fertile and suitable for potato cultivation. It can be observed that during harvesting seasons, the production of potatoes over demand brings down its price. This leads to the loss of potato farmers as the elasticity of demand for potatoes, like other staple crops, could be higher. During the off season, the price of potatoes entirely depends on the arrival of the market, as the demand for potatoes does not change significantly. Therefore, the significant reasons behind the volatility of potato prices are mainly supply-side-based.

The previous decade has witnessed undesirable volatility in the price of potatoes in the district of Burdwan, West Bengal, India. The potato supply was significantly affected as the farmers and traders were perplexed by price volatility (Rezitis & Stavropoulos, 2009; Sckokai & Moro, 2009; Seale & Shonkwiler, 1987; Taya, 2012). For instance, on 24 November 2014, the potato price in Burdwan district was Rs. 1995/quintal; it fell to Rs. 389/quintal on 23rd March (Ghatak & Albayrak, 1994). The ambiguity caused by price volatility influences the decision-making process of cultivators or the producers of crops. This volatility in potato prices leads to heavy losses for the farmers, and these losses lead to the miserable condition of the small farmers whom the money lenders finance. Sometimes, they commit suicide to avoid the burden of debt and future uncertainty. This can be observed in the district of Burdwan district, where some farmers committed suicide in April 2015.

Therefore, it is necessary to look at the volatility of potato prices and their influencing factors and to analyze the decision-making process of potato farmers. The findings of this study will enable the policymakers to design policies to protect the interest of the potato farmers through proper planning and by choosing appropriate marketing strategies. The primary objective of this study is to examine the factors responsible for the volatility of potato prices and the decision-making process of potato farmers as a response to this price volatility. So far, less work has been done in this area. For the execution of this objective, we employ the generalized autoregressive conditional heteroscedasticity (GARCH), and we use the Ag market database (2021) (agmarknet.gov.in) provided by the Government of India (Economic Survey of India). The price considered for analysis in this study is the daily arrival price of potatoes in the market.

The remaining portion of this chapter has been designed as follows. In the second section, we explain the existing literature on the volatility of the price of agricultural commodities and potatoes. Section three discusses the data source and methodology. The fourth section presents the results of the estimation. Section five concludes the study.

Literature Review

The literature survey, in detail, guides us to understand existing scenarios in the considered field, methodologies used, procedures of estimations, data sources, eloquent explanation and reconciliation of the contradictory results. Besides that, the empirical literature review investigates the avenues for present and future research efforts associated with the subject topic. So, we can take advantage of previous researchers' knowledge through their published papers to find research gaps. It will also help us identify the factors which influence farmers' behaviours in their decision-making process and perception. The background literature of the current research is quite limited but has a long past, as follows.

Brorsen and Irwin (1996) studied agribusiness pricing and marketing strategy. The analysis addresses the development of crops sold using alternative marketing strategies, considering options for the future hedge, forward contract and futures options. However, the study concludes that the portfolio is out of balance. Gasson (1973) studied the motivational aspect of farmers' economic behaviour concerning available resources. Exploring the goals and values of farming occupation, the study underscores the facet of motivation. Values may refer to the social, intrinsic or expressive aspects that influence farmers' decision-making in situations of choice. The study concludes that farmers tend to have an implicit alignment to work. Shonkwiler and Emerson (1982) proposed the hypothesis that farmers, as rational economic agents, determine production. Anticipated import of tomatoes and other economic variables was considered in the study. The estimation of maximum likelihood was used to solve a simultaneous equation model, and the research outcome was consistent with the hypothesis of rational expectation.

Gillmor (1986) studied behavioural approaches to Irish farmers. The study focuses on the nature of farmers' decision-making. By conducting a survey, the study reveals that the farmer's decision is influenced by both the inherent and the instrumental objectives. Profitability was not a leading motivational factor, according to the survey. Between the two demography farmers, Meath farmers tended to be commercially motivated, while Leitrim farmers were least motivated. Zanias (1987) studied exports of tobacco from Greece. Adjustment costs were considered in ad hoc past analysis, and rational expectations were listed. The result was consistent with price competition; the difference was observed over a long and short period, and higher elasticity was found. Eales et al. (1990) investigated market prices and the individual expectations of farmers. The study claims that an estimation of individual probability would demonstrate overconfidence. Therefore, the study indicates that group judgement would be more accurate than personal evaluation, but that overconfidence cannot be removed.

Malcolm (1990) studied for the last 50 years (from 1940–1990) of farm management research and practice in Australia. Malcolm noticed that despite much research, farmers still depend on intuition, experience and simple budgeting, which is typical for most countries. The study reveals a limited value of the initially appealing methodological techniques. The study highlights the limitation of their model of reality following a rigorous and well-rounded 50-year survey of agricultural economics. Ajzen (1991) studied the theory of planned behaviour. Behavioural achievement depends on intention (motivation) and ability (behavioural control). The study undergoes an empirical discussion about the prediction of behaviour for intention and perceived behaviour control and correlates between them in various studies. The study further performs another empirical discussion prediction of intention from the attitude towards behaviour subjective norms and perceived behaviour control. The study concluded that the theory of planned behaviour deals with the complexities of human social behaviour. Incorporating the central concept in social and behavioural science, the study predicts the particular behaviour in specific control.

Beedell and Rehman (1999) studied the conservation behaviour of farmers using the Structural Social Psychology Methodology and Theory of Planned Behaviour (TPB). The study explains the managing process of holding landscape by the farmers. Hedge management behaviour is prominent among the farmers of Bedfordshire. The study concludes that farmers' perception of the conservation benefit of hedge management is good. Murray-Prior and Wright (2001) investigated farmers' responses to starting to change under uncertain conditions. The study shows that high-level techniques developed in response to uncertainty lead to delays in production response. The causes of lags are risk attitude, price expectation, cost of the transaction, cost of adjustment, etc. The study concludes that decisions are becoming inappropriate under uncertainty. Simplification of behavioural rules is an alternate solution.

Kahneman (2002) depicted intuitive judgement and choice. The study provides a view of three main subjects: heuristic judgement, risky selection and framing impacts. After a decade of socio-cognitive psychology analysis, the study observes intuitions, thoughts and preferences quickly and without much reflection comes to mind. The study concludes that the relationship between intuition and perception has been particularly fruitful in identifying ways in which intuitive thinking differs from intentional reasoning, and the notions of accessibility and dual-process analysis play a crucial role in various social and cognitive psychology fields. Acevedo and Krueger (2004) studied illusion and belief in personal relevance. The study reveals that two psychological processes contribute to belief: (1) illusion of a person and (2) belief in personal relevance. Using the multivariate analysis of variance (MANOVA), the study undergoes two studies with undergraduate samples and estimates the behaviour.

Edwards-Jones (2006) studied modelling farmer decision-making. The study reveals that farmers are interested in the government and the public as numerous factors influence their decisions beyond the farm boundary. Six types of factors influence farmers' decisions. These are socio-demographic factors, psychological factors, characteristics of the farm, the structure of the farm, social milieu and adaptation of innovation. Using the theory of planned behaviour, the study undergoes a quantitative method where correlation measures attitude and personality. The study concludes that attitudes are linked to behaviour and personal characteristics. Parker et al. (2007) observed that 'self-reported maximizers' are more likely to present problem-sensitive styles of decision-making, as evidenced by less cognitive coping self-reports, high dependence on others in decision-making, more significant decision-making avoidance and a growing tendency to regret. The results also revealed that the self-reported maximizers were more likely to experience spontaneous decision-making. Yet, the correlation between self-reported maximization and worse life consequences is not affected by different types of decision-making controls, willingness to make decisions and socio-economic variables.

Alevy et al. (2008) conducted a field experiment of anchoring. Anchoring is a cognitive bias influencing human behaviour, i.e. decisions are made based on recent observations and a fixed scale in mind. A pillar of cognitive studies is the development of choices during the selection process. A notable finding is the assessment and valuation of uninformative numerical 'anchors'. The study concludes that an individual's demand should reflect a maximum willingness to pay for units of a commodity. Sadler-Smith (2008) examined the function of intuition in collective learning. The study reveals that intuitions are decisions that have arrived fast, affectively charged and without conscious awareness. Intuition is a complex problem that has the power to influence and persuade, which develops shared meaning. The study explains a theoretical model focused on feed forward and feedback relations, intuition and interpretation. The study concludes that some influences of intuitions are difficult to capture.

Baker and Nofsinger (2010) studied heuristic behaviour. Heuristic involves the reduction of the cognitive resources needed to find a solution. Heuristic behaviour includes overconfidence, availability bias and representativeness, etc. Fiedler (2010) studied heuristic decision-making as a cognitive process. The study deals with three lottery outcomes, i.e. best outcome, worst outcome and worst-case probabilities. The study concludes that the overall selection rate, adequately anticipated by the priority heuristic, was close to chance, though the high agreement between the judges reflected a systematic response. Gigerenzer and Gaissmaier (2011) investigated heuristic decision-making. The study addresses the question that although heuristic saves effort, in an uncertain world, does heuristic cognitive function well? This study concludes that (a) individuals and organizations sometimes adaptively rely on simple heuristics, and (b) ignoring part of the information can lead to more accurate decisions than weighting and adding all the information, such as low predictability and small samples.

Strough et al. (2011) discussed a conceptual and inspiring paradigm of bias in decision-making throughout our lives. The study concentrates on skills and abilities correlating pragmatic, analytical and practical decision-making procedures. The study concludes that life's development pathways differ according to the bias investigated. Fafchamps and Minten (2012) attempted to accumulate the effect of SMS-based agrarian information on 1,000 farmers in 100 villages in Pune, Nashik, Ahmednagar, Dhule and Latur in Maharashtra. The study estimates the significant advantages of weather and market information provided by a commercial service called Reuters Market Light (RMLb) to Indian farmers on mobile phones. The study reveals that treatment has affected spatial arbitration and the grading of crops, but these effects are minor. No significant statistical effects on farmers' earnings were reported in the analysis, such as added crop value, loss of crops due to heavy rain or the likelihood of changes in cultivation practices and varieties of crops.

The risk tolerance of French farmers is calculated in a study by Reynaud and Couture (2012). Using a field experiment, the study compared risk preference with non-expected utility and context-dependent. The study found that the risk attitude was correlated with the lottery option. Furthermore, using a psychometric questionnaire, the study found that risk attitudes are inconsistent and associated with investment and the recreational dimension. Liu (2013) analyzed the decision of Chinese farmers on whether to follow a biotechnological approach to GM Bt cotton regarding the farmers' attitude to risk. After a field survey, the study showed that those farmers are often more risk-averse; they adopted GM Bt cotton later. But because the farmers are overweight ness to the new technology, they have accepted it earlier and have acquired more income. The study concludes that farmers' perceived risk impacts their decision to adapt to technology.

The relationship between farmers' attitudes to risk and investment opportunities has been analyzed by Maertens et al. (2013). This study focused on the farmer's assets and the results of a risk experiment. Comparing the willingness to pay for different yield distributions, this study observed a high incidence of risk-seeking behaviour among farmers, collecting information from three villages in Andhra Pradesh, namely Aurepalle, Kanzara and Kinked, with 246 farm household samples to build a risk-loving behaviour model. The study has found a strong correlation between risk-taking behaviour and wealth and assets. Also, farmers are prepared to take risks in cash crops but are opposed to food crops. This study is limited to the village level (only three villages), and the sample size is also small to explain the behavioural nature of farmers. Keshavarz and Karami (2014) analyzed farmers' decision-making process with a sample of 255 farmers in drought-prone areas. They exposed that farmers' decision to manage drought resulted from a complicated web of economic, natural, structural and cognitive factors. During the various phases of drought, different survival responses have been adopted, ranging from long-term and short-term adjustments. Although many farmers adopted necessary changes in response to drought, they needed help to overcome the adverse effects of drought. The study concludes that various economic subsidies, promoting affordable coping mechanisms and combining agricultural interventions with social support services are needed for farmers.

Dale (2015) studied the heuristics and biases. This study explores commonly used heuristics and their origins in psychology, and comprehending how heuristic work can give us a better knowledge of our biases and impacts leads to a better solution in the decision-making process. The study concludes that while heuristics can speed up our problems and decisions, errors and biased judgements can also be introduced. Holst et al. (2015) investigated the anchoring effect of agricultural goods on Dutch farmers and the evidence for anchoring found in the farmers' decision. The study reveals that, in a repeated experimental auction, there is a negative adjustment in the open bid or no change in the close bid, consisting of a sequence of four series auctions, 48 groups of five farmers each. The analysis also found the impact of previous bids on subsequent bids; validity was tested by modifying the probability of winning and amount.

The specific literature on commodity price volatility contains studies based on different commodities and country settings. Nijman and Beestma (1991) focused on inferences of a simple pricing model for a commodity regarding the marginal process of sugar futures prices. The finding of this study reveals that futures price prejudice depends linearly on conditional variance. Using the multivariate generalized autoregressive conditional heteroscedasticity (MGARCH) model, they observe a significant impact of the conditional variance on the change in futures prices obtained from monthly and daily MGARCH models. In another study, Yang et al. (1992) evaluated three alternative models – the GARCH model, the time trend variance model and an econometric model (heteroscedasticity). The non-nested experiment recommends that modelling the sources of heteroscedasticity is the ideal method because this removes heteroscedasticity in the sample wheat yields, and GARCH also corrects heteroscedasticity. Time trends alone may not specify the factual variance structure.

In their study, Ghatak and Albayrak (1994) measured the risk and uncertainty mainly due to market reasons. They observe the process by which land allocation could be affected by income. For studying the fluctuations in the prices of food, Shively (1996) used the autoregressive conditionally heteroscedastic (ARCH) model. The finding of this study reveals that to enhance market performance, the government may reinforce and increase cash crop production; in this work, we will use generalized autoregressive volatility to study the volatility of the price of Irish potatoes. Using monthly data from 2001 to 2006 and employing three seasonal indicators with ARCH and GARCH models, Ravindran et al. (2007) carried out the analysis of the variability of onion price, especially the volatility of price in the markets of Hyderabad, Karnataka, India. The seasonal index analysis of onion prices in the Hyderabad market reveals that the prices of the first and fourth quarters increased by 1% and 7%, respectively, compared to normal prices. The ARCH and GARCH analysis confirm that onion prices in the Hyderabad market are not persistent.

Using data collected from 1994–1995 to 2004–2005, Yogisha et al. (2007) showed mixed trends in the market for the arrival of potatoes and potato prices. They observe that seasonal price indices go up for potato, ragi, onion and nut immediately after harvesting, and the price indicators reach their highest level during the lean period. Hence, the market demand and supply should remain uniform to maintain uniform and average market prices. Padmashri (2009) examined the arrival of sunflowers and groundnuts in the Challakere and Sira markets and their price behaviour. The investigation demonstrates that the arrival and price of both crops of the Challakere market rely upon the price of one lag season. ARCH model is identified as the appropriate model to describe the variance as well as forecast the price of the arrival of sunflowers and groundnuts at the Challakere and Sira markets. Balcombe (2009) explored the persistence of commodity prices' volatilities from 1957 to 2009 and their possible factors. The study observed that volatilities of agricultural prices are persistent over time, and there exist substantial heterogeneities in terms of trends in price volatilities across the considered products.

In a study by Balcombe (2009), determinants and nature of volatility in agricultural prices were examined. The finding of this study revealed that the volatility transmission across price, oil price and exchange rate explains agricultural price volatility. Huchet-Bourdon (2011) pointed out that the prices and volatilities of eight selected agricultural products were lower in the late 2000s than in the 1970s. Jha (2012) critically investigated the existing state of agricultural and rural marketing, and the study traced the advancement of rural marketing from farming to rural marketing and further progress orientation and advocated moving towards the holistic kind of pastoral marketing orientation. In another study, Kashyap (2012) observed that rural markets seemed unaffected by the economic slowdown, leading to a complete turnaround in the marketing scenario. There has been a transformation in the rural infrastructure and economics, and he has tried to analyze the impact of the changes on rural income and consumption patterns.

By analyzing a monthly time series dataset of 20 years and by using an autoregressive heteroscedasticity model to study the volatility of Irish potatoes, Innocent et al. (2015) evaluated the effects of the reforms of the market on Irish potato price volatility in the Nyandarua district of Kenya. According to the findings of this study, an increase in production value leads to a reduction in the uncertainty and risk associated with price volatility. This background encourages conducting the present study.

Data and Methodology

This study draws on the Ag market data of the Government of India for the period 2003–2019. The data set consists of 4,840 daily observations of modal price (average of the minimum and maximum price in rupees per quintal). The potato price volatility in the Burdwan district, India, might be examined by employing the GARCH (p, q) model. Suitable statistical techniques are applied to the data to analyze data on market arrival and potato price movement. Using augmented dickey–fuller, stationary characteristics of the variables are tested (Dickey & Fuller, 1979). The general form of the GARCH (p, q) model can be written as:

$$\sigma_t^2 = \alpha_0 + \alpha_1 \gamma_{t-1}^2 + \alpha_2 \gamma_{t-2}^2 + \ldots + \alpha_p \gamma_{t-p}^2 + \beta_1 h_{t-1} + \beta_2 h_{t-2} + \ldots + \beta_q h_{t-q} + u_t \tag{7.1}$$

where $p > 0$; $q \geq 0$; $\alpha_0 > 0$; $\alpha_i \geq 0$ for $i = 1, 2, \ldots, p$; $\beta_y \geq 0$ for $y = 1, 2, \ldots, q$; and u_t is the error term. The value of 'p' indicates the lag-length of the residual returns, and 'q' measures the lag of the variances. In the present analysis, the lag-length is chosen on the basis of Akaike (1969) Information Criterion (AIC). The model verification comprises the test data for residual autocorrelation and heteroscedasticity of the regression equation.

Results and Discussion

The movement of the price of potatoes exhibits volatility clustering like other agricultural commodities (Panel A of Fig. 7.1). This means there are periods when these series show wide fluctuations over an extended period preceded by a relatively calm period. The rising inflationary price must be considered in the decision-making since it makes future financial planning difficult. Exporters/importers and producers (here farmers) are likely to be affected by the volatility of commodity prices. High volatility could mean huge losses or gains and, hence, more significant uncertainty.

Due to unexpected natural disasters, pests, diseases and frost, crop production has decreased suddenly, leading to high prices. These high prices encourage potato growers to bring under potatoes next year, resulting in an acute glut and deterioration in the market price. This cobweb-type demand is available in the country's potato supply chain, which leads to high volatility in potato prices and has severe political effects. For example, a significant reduction in potato production in West Bengal was seen in 2009. During 2009, potato output decreased to 40% due to late blight syndrome. Only 5.4 million tonnes were produced

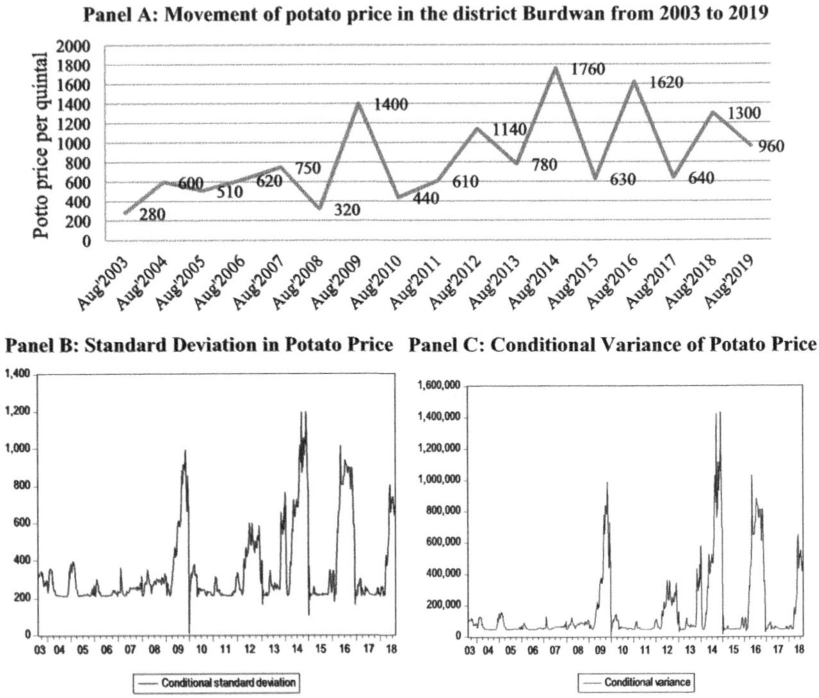

Fig. 7.1. Trends of Potato Price, Standard Deviation and Conditional Variance of Potato Price in the District Burdwan From 2003 to 2019. *Source:* Authors' own presentation using data from agmarknet.gov.in (2021).

against the previous year's production of 9.0 million tonnes, and as a result, the potato prices became high.

It emerges that daily percentage (converted to weekly) changes in commodity price exhibit volatility clustering, in which there are some periods of high volatility, such as between 2008–2009, 2014–2015 and 2015–2016, as shown in Panel B and C of Fig. 7.1.

$$\text{GARCH} = C(3) + C(4) \times \text{RESID}(-1)^2 + C(5) \times \text{GARCH}(-1)$$

The result of the unit root test is given in Panel I, and the estimated GARCH model is given in Panel II of Table 7.1. Panel II shows that the coefficients of the lag-squared residual and conditional variance of lag terms in the dependent variance equation are statistically significant (p-value = 0.000) at a 1% level. It indicates the presence of volatility clustering in the series of potato prices. Also, the total coefficients on the lag-squared residual (0.8300) and lag conditional variance (0.8161) are very close to one, implying that volatility shocks in the potato price are pretty persistent.

Table 7.1. Augmented Dickey–Fullar Test Statistics and Estimated Statistics of GARCH (1,1) Model.

Panel I: Augmented Dickey–Fullar Test Statistics

Series	*t*-Statistics	Prob.*
Test stastics with constant	−2.5038	0.0000
Test stastics with constant and trend	−2.3122	

Panel II: Estimated Statistics of GARCH (1,1) Model

Variable	Coefficient	Std. Error	*z*-Statistic	Prob.
C	−0.0031	0.0059	0.5144	0.0069
Potato	0.9297	0.0163	0.7733	0.0326
	Variance Equation			
C	7511.7712	4700.6722	1.4684	0.0058
RESID(−1)^2	0.8300	0.0563	3.3108	0.0000
GARCH(−1)	0.8161	0.0648	13.2699	0.0000
R^2	0.2117	Mean dependent var		0.0053
Adjusted R^2	0.1409	Standard dev. dependent var		0.0441
Standard error	0.0348	Akaike (1969) info criterion		−3.5669
Sum sq. residual	0.2073	Schwarz (1978) criterion		−3.3920
Log likelihood	197.8326	Hannan and Quinn (1979) criterion		−3.4960
Durbin and Watson (1950) stat	2.6005			

Source: Authors' own calculations using data from agmarknet.gov.in (2021).

Note: Akaike (1969) Information Criterion (AIC) is used for lag-length selection.
*Prob.** - significant results.

Concluding Remarks

GARCH is the most suitable process for forecasting the time series in volatility clustering in the financial market. The author proposed a GARCH (1, 1) model intending to forecast a short-term period of daily data on potato prices traded in the Burdwan market. This model is best suited in the short run as an instrument for risk management where the producers of potatoes might utilize the capability of prediction of the model. There has been a trend correction from 2003 to 2019, but in 2009 and 2015, potato price volatility peaked. The economic relevance of the result lies in its implications for potato production, especially for farmers. Price fluctuation is more persistent in the short run, thus making the model appropriate and accurate for predictions. The problem lies in the supply side, which indicates that producers cannot adjust themselves to the price volatility in the short run. Potato producers can go for contract farming or contract selling for the long run to protect the price of potatoes and nullify the price fluctuation effect.

This study has been based on secondary data collected from publications by government sources. Primary data may be beneficial for a better understanding of the arrival of potatoes and their price consistency. This study has been conducted in the Burdwan district of West Bengal. The market of varied demography, which includes other states, may be influenced by many other factors. Last but not least, further study can be carried out considering the potato prices in different regions of India and abroad.

References

Acevedo, M., & Krueger, J. I. (2004). Two egocentric sources of the decision to vote: The voter's illusion and the belief in personal relevance. *Political Psychology*, *25*(1), 115–134.

Ajzen, I. (1991). The theory of planned behavior. *Organizational Behavior and Human Decision Processes*, *50*(2), 179–211.

Akaike, H. (1969). Fitting autoregressive models for prediction. *Annals of the Institute of Statistical Mathematics*, *21*(1), 243–247.

Alevy, J., Landry, C., & List, J. (2008). *Anchors away: Field experiments on anchoring of economic valuations*. WorkingPaperecu0810 GreenvilleNCEast Carolina. https://economics.ecu.edu/wp-content/pv-uploads/sites/165/2019/07/ecu0810.pdf

Apergis, N., & Rezitis, A. N. (2011). Food price volatility and macroeconomic factors: Evidence from GARCH and GARCH-X estimates. *Journal of Agricultural and Applied Economics*, *50*(2), 179–211.

Baker, H. K., & Nofsinger, J. R. (Eds.). (2010). *Behavioral finance: Investors, corporations, and markets* (Vol. 6). John Wiley & Sons.

Balcombe, K. (2009). The nature and determinants of volatility in agricultural prices: An empirical study from 1962–2008. In A. Sarris & J. Morrison (Eds.), *The evolving structure of world agricultural trade*. Food and Agriculture Organization of the United Nations (FAO).

Beedell, J. D. C., & Rehman, T. (1999). Explaining farmers' conservation behaviour: Why do farmers behave the way they do? *Journal of Environmental Management*, *57*(3), 165–176.

Brorsen, B. W., & Irwin, S. H. (1996). Improving the relevance of research on price forecasting and marketing strategies. *Agricultural & Resource Economics Review*, *25*(1203-2016-94807), 68–75.

Dale, S. (2015). Heuristics and biases: The science of decision-making. *Business Information Review*, *32*(2), 93–99.

Degiannakis, S., Filis, G., Klein, T., & Walther, T. (2022). Forecasting realized volatility of agricultural commodities. *International Journal of Forecasting*, *38*(1), 74–96.

Dickey, D. A., & Fuller, W. A. (1979). Distribution of the estimators for autoregressive time series with a unit root. *Journal of the American Statistical Association*, *74*(366a), 427–431.

Durbin, J., & Watson, G. S. (1950). Testing for serial correlation in least squares regression: I. *Biometrika*, *37*(3/4), 409–428.

Eales, J. S., Engel, B. K., Hauser, R. J., & Thompson, S. R. (1990). Grain price expectations of Illinois farmers and grain merchandisers. *American Journal of Agricultural Economics*, *72*(3), 701–708.

Edwards-Jones, G. (2006). Modelling farmer decision-making: Concepts, progress and challenges. *Animal Science*, *82*(6), 783–790.

Fafchamps, M., & Minten, B. (2012). Impact of SMS-based agricultural information on Indian farmers. *The World Bank Economic Review*, *26*(3), 383–414.

Fiedler, K. (2010). How to study cognitive decision algorithms: The case of the priority heuristic. *Judgment and Decision Making*, *5*(1), 21.

Gasson, R. (1973). Goals and values of farmers. *Journal of Agricultural Economics*, *24*(3), 521–542.

Ghatak, S., & Albayrak, N. (1994). *Supply response in Turkish agriculture: An error correction and cointegration analysis: 1950–1990*. Discussion Papers in Economics-Department of Economics, University of Leicester.

Gigerenzer, G., & Gaissmaier, W. (2011). Heuristic decision making. *Annual Review of Psychology*, *62*, 451–482.

Gillmor, D. A. (1986). Behavioural studies in agriculture: Goals, values and enterprise choice. *Irish Journal of Agricultural Economics and Rural Sociology*, 19–33.

Hannan, E. J., & Quinn, B. G. (1979). The determination of the order of an autoregression. *Journal of the Royal Statistical Society: Series B*, *41*(2), 190–195.

Holst, G. S., Hermann, D., & Musshoff, O. (2015). Anchoring effects in an experimental auction – Are farmers anchored? *Journal of Economic Psychology*, *48*, 106–117.

Huchet-Bourdon, M. (2011). *Agricultural commodity price volatility: An overview*. Food, Agriculture and Fisheries Paper No. 52. OECD. https://doi.org/10.1787/18156797

Innocent, N., Mung'atu, J. K., & Thomas, K. (2015). Forecasting price of Irish potatoes volatility using GARCH and VECM models in Rwanda. *International Journal of Thesis Projects and Dissertations*, *3*(3), 44–53.

Jha, M. (2012). Holistic rural marketing. *International Journal of Rural Management*, *8*(1–2), 121–132.

Kahneman, D. (2002). Maps of bounded rationality: A perspective on intuitive judgment and choice. *Nobel Prize lecture*, *8*, 351–401.

Kashyap, P. (2012). The rural boom in India. *International Journal of Rural Management*, *8*(1–2), 133–141.

Keshavarz, M., & Karami, E. (2014). Farmers' decision-making process under drought. *Journal of Arid Environments*, *108*, 43–56.

Liu, E. M. (2013). Time to change what to sow: Risk preferences and technology adoption decisions of cotton farmers in China. *The Review of Economics and Statistics*, *95*(4), 1386–1403.

Maertens, A., Chari, A. V., & Just, D. R. (2013). Why farmers sometimes love risks: Evidence from India. *Economic Development and Cultural Change*, *62*(2), 239–274.

Malcolm, B. (1990). Fifty years of farm management in Australia: Survey and review. *Review of Marketing and Agricultural Economics*, *58*(430-2016-31539), 24–55.

Murray-Prior, R. B., & Wright, V. E. (2001). Influence of strategies and heuristics on farmers' response to change under uncertainty. *The Australian Journal of Agricultural and Resource Economics*, *45*(4), 573–598.

Nijman, T. E., & Beetsma, R. (1991). Empirical tests of a simple pricing model for sugar futures. *Annalesd' Economieet de Statistique*, 121–131.

Padmashri, H. S. (2009). *Time series analysis of arrivals and prices of groundnut and sunflower in Sira and Challakere market*. Doctoral dissertation. University of Agricultural Sciences GKVK.

Parker, A. M., De Bruin, W. B., & Fischhoff, B. (2007). Maximizers versus satisficers: Decision-making styles, competence, and outcomes. *Judgment and Decision making*, *2*(6), 342.

Ravindran, P. N., Babu, K. N., & Sivaraman, K. (2007). *Turmeric: The genus Curcuma*. CRC Press.

Reynaud, A., & Couture, S. (2012). Stability of risk preference measures: Results from a field experiment on French farmers. *Theory and Decision*, *73*(2), 203–221.

Rezitis, A. N., & Stavropoulos, K. S. (2009). Modeling pork supply response and price volatility: The case of Greece. *Journal of Agricultural & Applied Economics*, *41*(1), 145–162.

Sadler-Smith, E. (2008). The role of intuition in collective learning and the development of shared meaning. *Advances in Developing Human Resources*, *10*(4), 494–508.

Schwarz, G. (1978). Estimating the dimension of a model. *Annals of Statistics*, *6*(2), 461–464.

Sckokai, P., & Moro, D. (2009). Modelling the impact of the CAP single farm payment on farm investment and output. *European Review of Agricultural Economics*, *36*(3), 395–423.

Seale, J. L., & Shonkwiler, J. S. (1987). Rationality, price risk, and response. *Journal of Agricultural & Applied Economics*, *19*(1), 111–118.

Shively, G. E. (1996). Food price variability and economic reform: An ARCH approach for Ghana. *American Journal of Agricultural Economics*, *78*(1), 126–136.

Shonkwiler, J. S., & Emerson, R. D. (1982). Imports and the supply of winter tomatoes: An application of rational expectations. *American Journal of Agricultural Economics*, *64*(4), 634–641.

Strough, J., Karns, T. E., & Schlosnagle, L. (2011). Decision-making heuristics and biases across the life span. *Annals of the New York Academy of Sciences*, *1235*, 57.

Taya, S. (2012). *Stochastic model development and price volatility analysis*. OECD food, agricultural and fisheries working papers. No. 57. OECD publishing. http://dx.doi.org/10.1787/5k95tmlz3522-en. Accessed on October 2012.

Yang, S. R., Koo, W. W., & Wilson, W. W. (1992). Heteroskedasticity in crop yield models. *Journal of Agricultural and Resource Economics*, 103–109.

Yogisha, G. M., Karnool, N. N., & Kumar, H. S. (2007). Vijaya: Basavaraja, H. Trends and seasonal variations in arrivals and prices of potato in Kolar district. *Agricultural Marketing*, *49*(4), 26–28.

Zanias, G. P. (1987). Adjustment costs and rational expectations: An application to a tobacco export model. *American Journal of Agricultural Economics*, *69*(1), 22–29.

Chapter 8

Growth and Stability in Paddy Production and Irrigation Facilities: A State-Level Analysis in India With Special Reference to the Blocks of Paschim Medinipur District in West Bengal

Somdip Bhukta, Sunanda Das and Ramesh Chandra Das

Vidyasagar University, India

Abstract

The agricultural sector in any country plays an important contributor to national output along with its high absorbability of the working population. In the agricultural sector, paddy production holds a major share in the total output in countries like India. The growth and stability of the paddy production at the national level depends on the same factors at the sub-national levels such as the provincial and district levels. The present study focuses first on the stability analysis of the quantity of food production and irrigation facilities in the top 10 rice-producing states in India and then on the stability of yield rates in three varieties of paddy production, Aus, Aman, and Boro, and irrigation facilities across the blocks in the Paschim Medinipur district of West Bengal, as a special reference. The analyses involves the evaluation of the growth and stability using Coppock's method and trend analysis. It is found that Uttar Pradesh and Andhra Pradesh have a positive significant correlation between irrigation facilities and paddy production. From this study, it is also found that the correlation between changes in the yield rate of Aus, Aman, and Boro production and changes in irrigation is positive and significantly related to each other for Garbeta 1, Garbeta 3, Keshiary, Kharagpur 2, Chandrakona 1, and Chandrakona 2 blocks in Paschim Medinipur district under West Bengal. Therefore, the results recommend that the government should focus on irrigation facilities in those areas where the benefits of irrigation facilities are fully realized.

Growth, Poverty and Developmental Aspects of Agriculture, 107–119

Published under exclusive licence by Emerald Publishing Limited
doi:10.1108/978-1-83608-076-320241008

Keywords: Coppock's index; stability; agriculture; food production; paddy production; yield rates; Aus; Aman; Boro; irrigation

Introduction

India is well known for its agricultural production, which contributes approximately 18% of the gross domestic product and creates a large amount of employment in rural areas. Agricultural production cares about the livelihood of 70% of the rural population, and it also creates food security for the Indian economy. Agriculture is the backbone of the Indian economy because it develops the economy in many ways such as product contribution, availability of raw materials, market contribution, and factor contribution where product contribution is related to the availability of food, the market contribution is associated with goods produced for another sector, and factor contribution implying labor and capital availability of agricultural sector. On the other side, the production of paddy and rice is an important part of the national economy, and rice is a basic food crop in the tropical region. Rice is produced in assured irrigated and rainfed areas, and rice can be produced in both the Kharif and Rabi seasons. India is the second-largest producer of rice; the rice are mainly of three types: Aus, Aman, and Boro. However, rice production in India mainly depends on rainfall, which creates instability in the rice production in India over the years due to vagaries of nature. This instability in rice production influences the price stability and consumer standard of living and also increases the vulnerability of low-income households of the country, states, and regions. Therefore, stability in rice production is important for the economic stability. On the other hand, in order to mitigate the fluctuations in the rainfall as well as to reduce the production instability in paddy or rice production, investment in the irrigation sector, modernization of irrigation techniques, and improved and sustainable water management practices is needed. Irrigation facilities in India have the sufficient capacity of increasing total factor productivity growth in India in its agricultural sector (Das, 2023). Irrigation, as part of public capital formation, influences the private capital formation leading to forward and backward linkage effects between agriculture and institutional factors. The phenomenon of such linkage effects has been the outcome of many research outputs such as of Aschauer (1989), Erden and Holcombe (2006), Furceri and Sousa (2009), Şen and Kaya (2014), and Das et al. (2018).

In addition to the aforementioned natural problems, the agricultural sector has been suffering from uncertainties, risks, and problems including environment-related risks, market-related risks, and financial risks, thereby making agricultural production unstable. There are different types of risk: first is market-related risk, which is mainly related to price volatility, and price volatility is related to demand and supply of the food grain. It is also well known that demand for agricultural products is generally price inelastic, i.e., little increase in supply will lead to a sharp decline in the price. Another risk is the environment or climate change that can create instability in rice production. Lastly, financial risk is related to credit given to the farmer; if this risk is high, the farmer will not use the input optimally, which can also create instability in the rice production. Although India is second largest producer of rice, rice production in India most of the time follows the traditional path, which is associated with risks and thus makes rice production unstable. The study by Kumar and Jain (2013) has observed a large variation in the productivity of the crop within the states in India and the country as a whole. In another study, it was found that the instability of agricultural production had been amplified with the adaptation of new technology (Mehra, 1981). On the other side, Dev (1987) found a progressive but marginal decline in instability in food grain production at the national level. It was noticed in another study that there was no positive relation between growth and instability (Goyari, 2013); it was discovered that weather uncertainty was only the responsible factor for higher instability. Therefore, instability affects production and productivity, which makes the growth of the agricultural sector unstable.

In this context, it is important to understand through a fresh investigation how instability in paddy production is related to instability in the irrigation facilities. To fulfill this purpose, we have taken the top 10 paddy-producing states in India. Moreover, to get a deeper understanding on the depth of the instability problem, we carried out a micro-level study. We have thus taken the block-level data of Aus, Aman, and Boro in Paschim Medinipur district of the West Bengal state. The production of Aus, Aman, and Boro is one of the important crops among agricultural products; therefore, there is a need to understand the instability of rice production over the years.

Literature Review

This section highlights some studies available in the existing literature which are related to the present study. Pillai (2001) analyzed the growth of paddy production in West Bengal and Orissa and found that total factor productivity was positive for both states, but the performance was better for West Bengal because farmers in West Bengal utilized efficient techniques like using a variety of seeds for different seasons. Swain (2014) analyzed the instability of agricultural production in Western Odisha. The yield decomposition analysis was used to examine the role of risk factors for the change in the productivity of inputs in the crop yield. It was observed that 80% change in productivity is due to factors such as rainfall failure, rainfall variability, drought, and high temperature, and another 20% change is due to factors such as labor, fertilizer, pesticides, and irrigation. Sable et al. (2020) examined how Indo Gangetic basin is responsible for the production of seed and paddy. Using multivariate regression analysis, liner log, and Cobb–Douglas production function, they found that the Nadia district is good for seed production for in both Kharif and Rabi seasons. Jain (2018) studied the instability of different agricultural dimensions such as area, production, yield, and price. Using the compound annual growth rate, it was found that the production was positive, but it declined over the period. The researcher used the standard deviation of the natural log as the instability index. Mandal et al. (2018) broke down the farmers' practice for rice cultivation under the red-laterite zone.

They carried out a production-oriented survey to find out the results. It is found that the farmer does not follow the fertilizer schedule as per the recommendation, but the cultivator in this zone adopted need-based plant protection measures. Sekhara and Devarajulu (2019) analyzed the trend of the area, production, and productivity of paddy using compound annual growth rate and t-statistic. In this study, the researcher suggested some policies to improve productivity in agriculture. The existence of forward and backward linkage within the rural sector and with the strategy of overall economic growth and development was observed. Rahman and Parkinson (2007) analyzed how productivity and soil fertility are related to rice production. It was observed that soil fertility had a significant impact on rice production. It was also discovered that inherent knowledge transfer was also the responsible factor for higher productivity. Gangwar and Katyal (2001) examined the productivity, stability, and profitability of rice-based crop sequences in West Bengal. It was observed that the rice-potato-jute sequence gives high profit in Kalyani. At the same time, the stability of the rice-potato-jute sequence was higher in Kalyani. Mukherjee and Huda (2018) analyzed on how climate variability affects the productivity trend in West Bengal. The researchers used a crop growth simulation approach and discovered that productivity improvement was possible if farmers adopted temperature and stress management techniques and better management procedures. On the other hand, Aggarwal et al. (2000) broke down rice-wheat trend system in northwestern India. The researcher used a simulation model and noted a significant decline when the previous yield was high. It was also noticed that high temperature was an important factor stagnation of agricultural products. Gunri et al. (2004) analyzed how integrated nitrogen application and spacing affect rice production. From the study, it was revealed that the efficiency of nitrogen utilization was higher for green manure. It was discovered that nitrogen recovery was higher for the local level of nitrogen, and it decreased significantly for higher levels of nitrogen. Ray et al. (2018) analyzed how technology can influence the economy and the environment by considering the rice-based cropping system in India. They used the analysis of variance (ANOVA) test, rice equivalent yield system, and global warming potential parameter and discovered that a rice-based cropping system could positively influence profit and energy use. Manjunath et al. (2018) examined the rice-based cropping system in Goa and West Bengal by using ANOVA and cost–benefit ratio. It was found that sweet corn production was energy efficient and gave the highest return. Duxbury (2001) analyzed the trend of the rice-wheat cropping system. The researcher used a simple regression analysis and found that various organic materials are partially substituted for the fertilizer nitrogen, phosphorus, and potassium. Chaudhary et al. (2019) examined the higher resource practices for higher resource efficiency and productivity. It was noticed that nitrogen fertilizer should be given before the irrigation to get the best effect of nitrogen. In addition to this, Roy et al. (2011) studied the sustainability of the cropping system of lowlands in north Bihar. To find out the result, the researcher used the sustainability yield index. It was found that rice-lentil-green gram had the highest yield and highest return. Along with this, Singh et al. (2012) investigated how a diversified cropping system was helpful for sustainability, productivity, and profitability. They used the diversity index to measure the diversity and discovered that the diversifying index was higher for small farmers than for medium and large farmers. Das (2023) computed total factor productivity growth (TFPG) out of the institutional roles through public irrigation facilities and commercial bank credit besides traditional inputs in Indian agricultural sector for the post-globalized era of 1991–2019 in Indian agriculture sector. Using the growth accounting approach, the study found that the institutional factors, such as public irrigation facilities and bank finances, have contributed significantly to the growth of the agricultural output in India and the values of TFPG have increased over time.

Gaps in the Literature

Many studies have established how rice production is affected by various factors such as soil fertility, irrigation, weather conditions, management technique, crop sequence, and fertilizer; however, there are no studies that describe the relationship between the instability of the paddy production at the state levels in general and instability of irrigation facilities in the production of Aus, Aman, and Boro at block levels. Therefore, to fill this gap, this study would like to discuss the instability of paddy production and instability in irrigation of the top 10 rice-producing states in India; in addition, the study would like to discuss a block-level analysis to know the instability in rice production at block level.

Objectives of the Study

- To measure the instability of rice production across the top 10 rice-producing states in India.
- To analyze the instability in rice production (Aus, Aman, and Boro) and irrigation across the blocks in the Paschim Medinipur district.
- To analyze the correlation between changes in irrigation and changes in rice production across the top 10 rice-producing states in India.
- To analyze the correlation between changes in irrigation and changes in rice production across the blocks in the Paschim Medinipur district.

Data Source

The yield rate data of Aus, Aman, Boro, and irrigation data are available in the district statistical handbook. All the data are time series data available from 2002 to 2016. Total food grain data of the top 10 rice-producing states are available in the *RBI Statistical Handbook* (www.rbi.org). Data have been taken from 2003 to 2023. Irrigation Potential created under the Accelerated Irrigation Benefit Programme (AIBP) is collected from agricultural statistics at a glance (2019). Rice production is highest in West Bengal but in terms of stability, West Bengal takes second place among the top 10 rice-producing states in India. In addition to this, rice production in Paschim Medinipur in West Bengal has been moderate to high (Biswas, 2022) from 2001 to 2015 which is the reason for taking all blocks in

Paschim Medinipur district to measure the instability of rice production across the blocks.

Here, the study uses the Coppock's index to measure instability and then finds the correlation between changes in irrigation and food grain production of the top 10 rice-producing states in India. The study will also measure the instability of block-level data of yield rates in paddy production in Paschim Medinipur district under West Bengal.

Methodology

We use the well-known Coppock's index to measure the instability.

The formula of the Coppock's instability index is given below:

$$\text{Coppock instability index} = \text{antiolog}\left(\sqrt{\text{Vlog}} - 1\right)$$

where $\text{Vlog} = \sum\left(\log\frac{x_{t+1}}{x_t} - m\right)^2 / n$.

Here, x_t = yield rate and irrigation rate at time t.

t = number of years.

m = mean of the difference between logs of x_t, x_{t+1}.

V log = logarithmic variance of the series.

We carried out the correlation analysis between the changes in irrigation and the yield rates of three crops by Pearson's product moment formula. Then, we go for the t-test of such correlation coefficients.

Changes in irrigation is denoted by ΔI, and changes in production is denoted by ΔP.

Here, $\Delta I = \left(\log\frac{i_{t+1}}{i_t} - m\right)^2$ and $\Delta P = \left(\log\frac{p_{t+1}}{p_t} - m\right)^2$.

i_t and p_t indicate irrigation and production, respectively, at the time t.

i_{t+1} and p_{t+1} indicate irrigation and production, respectively, in the time $t + 1$.

The formula to calculate the t-score is:

$$t = r\sqrt{\left(\frac{n-2}{1-r^2}\right)}$$

where "r" is the coefficient of correlation and "n" is the sample size.

Results and Discussion

We have calculated the instability index values of the top 10 rice-producing states using the aforementioned formula, and the results are presented in Table 8.1 in Column 2 and sub-columns of Column 2 and 3. The table also depicts the correlation results in Column 3. It is found that Punjab has the lowest instability in total food grain production, followed by West Bengal, Odisha, Uttar Pradesh, and Haryana.

Table 8.1. Values of Coppock Index of the Top 10 Rice-Producing States in India and All the Blocks of Paschim Medinipur District in West Bengal.

Coppock's Index Values of Top 10 Rice-Producing States in India				
West Bengal	38.2		−0.3 (−1.0)	
Uttar Pradesh	38.9		0.7 (3.4)**	
Punjab	37.9		−0.2 (−0.9)	
Tamil Nadu	42.3		0.0 (0.0)	
Andhra Pradesh	40.4		0.6 (2.5)**	
Bihar	40.8		0.0 (−0.1)	
Chhattisgarh	39.8		0.3 (1.0)	
Odisha	38.9		−0.2 (−0.6)	
Assam	39.7		0.2 (−0.9)	
Haryana	38.5		−0.2 (−0.7)	
Coppock's Index Values of All the Blocks of Paschim Medinipur District				
	Aus	Aman	Boro	Irrigation
Jhargram	94.19	56.82	61.66	39.70
Binpur 1	100.31	56.29	70.93	45.13
Binpur 2	92.13	55.53	102.65	44.62
Jamboni	57.08	64.68	81.54	45.87
Nayagram	133.91	63.3	62.87	40.59
Sankrail	54.36	59.39	52.50	42.93
Gopiballavpur 1	64.30	65.18	61.13	47.33
Gopiballavpur 2	60.84	55.02	61.86	46.89
Salboni	56.09	56.08	57.15	44.01
Keshpur	73.66	67.70	49.94	48.51
Garbeta 1	63.02	53.69	57.46	42.24
Garbeta 2	91.27	54.40	63.56	42.63
Garbeta 3	49.84	58.99	61.80	42.71
Midnapore	70.51	59.06	65.92	41.90
Debra	79.40	63.67	77.86	41.34
Pingla	66.87	57.55	54.88	38.99
Keshiary	64.61	59.65	58.64	46.17
Dantan 1	97.62	58.75	50.82	40.48
Dantan 2	90.18	57.99	48.95	42.88
Narayangarh	95.28	59.69	56.43	41.64
Mohanpur	101.4	59.89	49.69	40.08

(Continued)

Table 8.1. *(Continued)*

Sobong	66.94	58.46	65.51	39.34
Kharagpur 1	111.24	60.62	73.59	41.51
Kharagpur 2	61.73	63.06	61.87	49.88
Chandrakona 1	80.17	66.37	53.19	41.46
Chandrakona 2	99.26	65.03	49.02	45.33
Ghatal	69.09	62.53	56.26	37.71
Daspur 1	109.20	70.47	59.28	39.88
Daspur 2	63.18	131.30	58.84	38.31
Correlation and T-test values along the cross section				
Correlation (Aus and irrigation)	−0.28 (−1.52)			
Correlation (Aman and irrigation)	−0.23 (−1.22)			
Correlation (Aman and irrigation)	0.13 (0.68)			

Source: Authors' calculations.

Note: "**" → Significant at 5 % level. Column 3 → Correlation between changes in production and changes in irrigation.

On the other side, Tamil Nadu shows the highest instability in total food grain production; one of the reasons of this phenomenon is zero potential created under the Accelerated Irrigation Benefit Program (Agricultural Statistics at a Glance, 2019). The study has also found a significant positive correlation between change in food production and change in irrigation in Uttar Pradesh and Andhra Pradesh, as the potential created for irrigation is higher for these two states than other rice-producing states in India. In addition to this, rice production is highest in West Bengal, but in terms of stability, West Bengal takes second place among the top 10 rice-producing states in India. In addition to this, rice production in the Paschim Medinipur in West Bengal is moderate to high (Biswas, 2022) from 2001 to 2015, which is the reason for taking all blocks in Paschim Medinipur district to measure the instability of rice production across the blocks.

Here, the study has derived the instability of four items, namely Aus, Aman, Boro, and irrigation across the different blocks in Paschim Medinipur district. It is found that magnitudes of instability are higher for Aus production in the districts like Jhargram, Binpur 1, Nayagram, Gopiballavpur 2, Garbeta 1, Garbeta 2, Midnapur, Debra, Pingla, Keshiary, Dantan 1, Dantan 2, Narayangarh, Mohanpur. Sobong, Kharagpur 1, Chandrakona 1, Chandrakona 2, Ghatal, and Daspur 1. On the other side, Sankrail, Gopiballavpur 1, Kharagpur 2, and Daspur 2 have found higher instability in Aman production among all three crops. Moreover, Binpur, Jamboni, Salboni, Keshpur, and Garbeta 3 have found greater instability in Boro production. Now, we will compare the instability across the blocks. The instability of Aman production is highest for Nayagram, followed by Kharagpur block across all the blocks in the district. At the same time, the instability index in Aus production is lowest for the Garbeta block. The instability for Aman production is highest for Daspur 1 block and it is lowest for Garbeta 1 block. On the other side, the instability for Boro production is higher for the Binpur 2 block, followed by Jamboni and Binpur 1, and the instability of Boro production is lowest for Datan 2, Keshpur, Mohanpur, and Chandrakona 2. In addition to this, the instability of irrigation is highest for Kharagpur 2 block, and the instability of irrigation is lowest for Ghatal, followed by Daspur 1 and Daspur 2.

After the computation of the instability index, we would like to discuss the instability of rice production and irrigation. The result is insignificant, but the sign is expected, i.e., the correlation between Boro and irrigation instability is positive, implying that irrigation may have some positive impact on Boro production, and other correlations found in Table 8.1 are negative. Now, the study will move toward discussing the result of Table 8.2.

In Table 8.2, we compute the correlation between $\frac{x_{t+1}}{x_t}$ and $\frac{i_{t+1}}{i_t}$ across the years for each of the blocks. A positive significant correlation between changes in irrigation and changes in Boro paddy production is found in Garbeta 1, Garbeta 2, and Kharagpur 2; however, Garbeta 1 block is significant at a 10% significance level. In addition, the correlation between changes in irrigation and changes in Aman production is positive and significant for Garbeta 3 and Chandrakona 1. Furthermore, the correlation between irrigation and yield rates in Aus is positive and significant for Garbeta 1, Keshiary, Dantan 1, and Daspur 1; however, Daspur 1 is significant at a 10% level. Hence, irrigational facilities, as a public institutional support, have close ties with the rice production in the states in India as well as in the micro units, the blocks, in the Paschim Medinipur district in West Bengal. Further steady intervention of the government sector in terms of providing more irrigational facilities, thus, could improve the production and productivity in the agriculture sector in the country as well as to reduce their volatilities.

Table 8.2. Correlation Between Changes in Yield Rates and Irrigation Facilities in the Three Crops Over the Years.

	Irrigation and Boro	**Irrigation and Aman**	**Irrigation and Aus**
Jhargram	0.4 (−1.4)	−0.2 (−0.6)	−0.2 (−0.8)
Binpur 1	−0.1 (−0.4)	−0.1 (−0.4)	−0.2 (−0.8)
Binpur 2	0.3 (−1.1)	−0.1 (−0.2)	−0.3 (−1.0)
Jamboni	0.3 (−1.3)	0.1 (0.4)	0.0 (0.1)
Nayagram	0.2 (0.6)	−0.1 (−0.4)	−0.2 (−0.7)
Sankrail	0.3 (−1.1)	−0.2 (−0.7)	−0.1 (−0.2)
Gopiballavpur 1	−0.2 (−0.6)	0.1 (0.4)	−0.1 (−0.4)
Gopiballavpur 2	0.2 (0.7)	−0.1 (−0.4)	−0.2 (−0.7)

(Continued)

Table 8.2. *(Continued)*

	Irrigation and Boro	Irrigation and Aman	Irrigation and Aus
Salboni	0.1 (−0.5)	−0.2 (−0.7)	−0.2 (−0.6)
Keshpur	0.2 (−0.7)	−0.3 (−1.1)	−0.3 (−1.2)
Garbeta 1	0.5 (1.8)*	0.4 (1.3)	0.5 (2.0)**
Garbeta 2	0.0 (0.0)	−0.1 (−0.4)	0.2 (0.7)
Garbeta 3	0.5 (2.0)**	0.5 (2.2)**	0.4 (1.5)
Midnapore	0.2 (−0.7)	0.0 (0.0)	−0.3 (−1.1)
Debra	0.2 (−0.6)	0.2 (0.7)	0.2 (0.7)
Pingla	0.2 (−0.5)	−0.2 (−0.6)	−0.1 (−0.4)
Keshiary	0.2 (−0.7)	−0.1 (−0.5)	0.7 (3.8)**
Dantan 1	−0.2 (−0.6)	0.1 (0.2)	0.6 (2.9)**
Dantan 2	−0.2 (−0.7)	0.0 (−0.1)	−0.1 (−0.2)
Narayangarh	0.2 (−0.5)	−0.2 (−0.5)	−0.2 (−0.6)
Mohanpur	0.1 (−0.3)	−0.2 (−0.6)	−0.2 (−0.8)
Sobong	0.2 (−0.6)	−0.1 (−0.4)	0.3 (0.9)
Khargapur 1	0.2 (0.6)	−0.2 (−0.6)	−0.1 (−0.4)
Khargapur 2	0.5 (2.0)**	0.3 (1.2)	0.0 (0.0)
Chandrokona 1	−0.1 (−0.5)	0.5 (2.0)**	−0.2 (−0.8)
Chandrokona 2	0.1 (0.3)	0.2 (0.8)	0.8 (5.6)**
Ghatal	0.2 (−0.8)	0.0 (−0.1)	−0.2 (−0.8)
Daspur 1	−0.1 (−0.5)	0.0 (0.2)	0.5 (1.8)*
Daspur 2	0.0 (−0.1)	0.0 (−0.1)	0.1 (0.2)

Source: Authors' calculations.

Note: "**" → Significant at 5% level, "*" → Significant at 10% level.

Conclusion

The agricultural production in India depends on various factors such as temperature, soil fertility, and irrigation. Instability in all these factors can create instability in agricultural production; for example, instability in the irrigation sector may be the factor for creating the instability in paddy production. The study thus has gone through measuring the instability in the paddy/rice production as well as yield rates using the Coppock's index across listed the top 10 rice-producing states in India and blocks under the Paschim Medinipur district in West Bengal.

In terms of stability in rice production, Punjab and West Bengal take first and second position, respectively. However, in terms of quantity of rice production,

West Bengal stands at the first place though potential created irrigation nearly zero from 2014 to 2018, which may create instability in rice production in the long term; as a result, West Bengal cannot hold its first position. According to the Business Line, West Bengal takes fourth position in rice production (Mishra, 2024).

On the other side, under the Accelerated Irrigation Benefit Programme (AIBP), potential irrigated land is close to zero for a long time in Tamil Nadu, which might create instability in rice production. From the result, one can observe that Tamil Nadu takes the last position in terms of stability in food grain production among the top 10 rice-producing states in India. On the contrary, Punjab takes the first position in the stability of food grain production, but the creation of irrigated areas under the AIBP is lower than Uttar Pradesh, Andhra Pradesh, Chhattisgarh, and Odisha. Therefore, it can be said that there are the other factors along with irrigation that may create stability in food production in Punjab.

The values of instability are close to each other, and the instability of Punjab is lower among all the top 10 rice-producing states. On the other side, West Bengal secures the first position in rice production, and Paschim Medinipur district has a vital role for giving a significant portion of rice production to the state economy of West Bengal. Therefore, we have taken all the blocks in this district to know the relationship between the changes in irrigation with the yield rates in paddy production. It is found that changes in irrigation are not significantly correlated with paddy production in most of the blocks in Paschim Medinipur; however, for some blocks, the change in irrigation has a positive impact on the change in rice production.

Limitations of the Study

Along with irrigation, there are other factors that create stability as well as instability in agricultural production. One can analyze the effect of those factors on the creation of stability in food grain production. In addition to this, one can use another method of instability measurement which might give different results. Moreover, this study can be extended to the state-level, district-level, and block-level analysis to get deeper intuition.

Future Scope of the Study

From the study, it is clear that irrigation has a relationship with rice and food grain production for some states and irrigation also has a relationship with rice production in some blocks at Paschim Medinipur district in West Bengal; however, for the other states and blocks mentioned in the study, such type of relationship is not found. Therefore, policymakers should improve the irrigation areas where it has an impact. For other blocks, policymakers should give attention to the other factors along with irrigation, which may create stability in agricultural production.

References

Aggarwal, P. K., Bandyopadhyay, S., Pathak, H., Kalra, N., Chander, S., & Kumar, S. (2000). Analysis of yield trends of the rice-wheat system in north-western India. *Outlook on Agriculture*, *29*(4), 259–268. https://doi.org/10.5367/000000000101293329

Agricultural Statistics. (2020). *Agricultural statistics at a glance 2019*. Official website of Directorate of Economics and Statistics, Department of Agriculture and Farmers Welfare, Ministry of Agriculture and Farmers Welfare, Government of India. Ministry of Agriculture and Farmers Welfare Department of Agriculture, Statistics. https://desagri.gov.in/document-report/agricultural-statistics-at-a-glance-2019/

Aschauer, D. (1989). Is public expenditure productive? *Journal of Monetary Economics*, *23*(2), 177–200.

Biswas, B. (2022). Changing crop concentration and agricultural efficiency: A study in West Bengal, India. *Geojournal*, *87*(2), 491–513.

Chaudhary, A., Chhokar, R. S., Yadav, D. B., Sindhu, V. K., Ram, H., Rawal, S., & Gill, S. C. (2019). In-situ paddy straw management practices for higher resource use efficiency and crop productivity in Indo-Gangetic Plains (IGP) of India. *Journal of Cereal Research*, *11*, 172–198.

Das, R. C. (2023). Sustainability through total factor productivity growth in agriculture incorporating institutional factors: A post-globalized Indian scenario. *International Journal of Social Ecology and Sustainable Development*, *14*(1), 1–16.

Das, R. C., Das, A., & Ray, K. (2018). Examining forward and backward linkages between public and private investments: A cross-country analysis. *Review of Market Integration*, *10*(1), 1–31.

Duxbury, J. M. (2001). Long-term yield trends in the rice-wheat cropping system. *Journal of Crop Production*, *3*(2), 27–52. https://doi.org/10.1300/j144v03n02_02

Erden, L., & Holcombe, R. G. (2006). The linkage between public and private investment: A cointegration analysis of a panel of developing countries. *Eastern Economic Journal*, *32*(3), 479–492.

Furceri, D., & Sousa, R. (2009). *The impact of government spending on the private sector: Crowding out versus crowding-in effects*. NIPE Working Papers 6. https://ssrn.com/abstract=1346329

Gangwar, B., & Katyal, V. (2001). Productivity, stability and profitability of rice (Oryza sativa)-based crop sequences in West Bengal and Orissa. *Indian Journal of Agronomy*, *46*(3), 387–394.

Goyari, K. R. (2013). Analyzing growth and instability in subsistence agriculture of Odisha: Evidence from major crops. *Agricultural Economics Research Review*, *26*, 67–78.

Gunri, S. K., Pal, S. K., & Choudhury, A. (2004). Effect of integrated nitrogen application and spacing on yield of rice (Oryza sativa) in foot-hill soils of West Bengal. *Indian Journal of Agronomy*, *49*(4), 248–250.

Jain, A. (2018). Analysis of growth and instability in area, production, yield and price of rice in India. *Social change and development*, *15*(2), 46–66.

Kumar, A., & Jain, R. (2013). Growth and instability in agricultural productivity: A district level analysis. *Agricultural Economics Research Review*, *26*(conf), 31–42.

Mahendradev, S. (1987). Growth and instability in foodgrains production: An inter-state analysis. *Economic and Political Weekly*, *22*(39), A82–A.

Mandal, R., Jana, K., Ghosh, S., Biswas, A., Bhadra, K. K., & Mallick, G. K. (2018). Production oriented survey (POS) on different aspects of rice cultivation and farmers practices under red and lateritic zone of West Bengal, India. *Archives of Agriculture and Environmental Science*, *3*(4), 378–381. https://doi.org/10.26832/24566632.2018.030408

Manjunath, B. L., Paramesh, V., Mahajan, G. R., Reddy, K. V., Das, B., & Singh, N. P. (2018). A five years study on the selection of rice-based cropping systems in Goa, for west coast region of India. *Journal of Environmental Biology*, *39*(3), 393–399. https://doi.org/10.22438/jeb/39/3/mrn-593

Mehra, S. (1981). *Instability in Indian agriculture in the context of new technology*. IFPRI.

Mishra, P. (2024, March 11). Fall in West Bengal output drags India's rice production. *BusinessLine*. https://www.thehindubusinessline.com/economy/agri-business/fall-in-west-bengal-output-drags-indias-rice-production/article67939499.ece. Accessed on March 20, 2024.

Mukherjee, A., & Huda, A. K. S. (2018). Assessment of climate variability and trend on wheat productivity in West Bengal, India: Crop growth simulation approach. *Climatic Change*, *147*(1–2), 235–252.

Pillai, R. (2001). An analysis of paddy productivity growth in West Bengal and Orissa. *Indian Journal of Agricultural Economics*, *56*(4), 613–630.

Rahman, S., & Parkinson, R. J. (2007). Productivity and soil fertility relationships in rice production systems, Bangladesh. *Agricultural Systems*, *92*(1–3), 318–333.

Ray, K., Hasan, S. S., & Goswami, R. (2018). Techno-economic and environmental assessment of different rice-based cropping systems in an inceptisol of West Bengal, India. *Journal of Cleaner Production*, *205*(2018), 350–363. https://doi.org/10.1016/j.jclepro.2018.09.037

Roy, D. K., Kumar, R., & Kumar, A. (2011). Production potentiality and sustainability of rice-based cropping sequences in flood prone lowlands of North Bihar. *ORYZA-An International Journal on Rice*, *48*(1), 47–51.

Sable, M. S., Nandi, A. K., & Biswas, R. (2020). Factors of paddy grain and seed production at farm level in the Indo-Gangetic basin of West Bengal. *Journal of Crop and Weed*, *16*(3), 97–103. https://doi.org/10.22271/09746315.2020.v16.i3.1372

Sekhara, K., & Devarajulu, M. (2019). Trends in area, production and productivity of paddy crop: An overview. *International Journal of Humanities and Social Science Invention*, *8*(1), 50–58.

Şen, H., & Kaya, A. (2014). Crowding-out or crowding-in? Analyzing the effects of government spending on private investment in Turkey. *Panoeconomicus*, *61*(6), 631–651.

Singh, R. D., Shivani, Khan, A. R., & Chandra, N. (2012). Sustainable productivity and profitability of diversified rice-based cropping systems in an irrigated ecosystem. *Archives of Agronomy and Soil Science*, *58*(8), 859–869. https://doi.org/10.1080/03650340.2011.554403

Swain, M. (2014). Sources of growth and instability in agricultural production in Western Odisha, India. *Asian Journal of Agriculture and Development*, *11*(1362-2016-107732), 51–70.

Section II

Poverty, Food Security and Developmental Aspects of Agriculture Across the Countries and Groups

Chapter 9

Unemployment, Food Insecurity and the Misery of Flood-Affected People in Bangladesh: How Do the Government and NGOs Respond?

*Tulshi Kumar Das and Md Mahmud Hasan**

Shahjalal University of Science and Technology, Bangladesh

Abstract

Despite recent economic progress, Bangladesh still faces significant challenges from natural disasters that impact food security and employment. In 2017, a severe flash flood caused by excessive rain and onrush from India devastated the *haor*-based Sunamganj district. The study explores the impacts of the flash flood on food security, employment and the overall well-being of the affected population, as well as evaluates government and NGO initiatives taken to alleviate the crisis using a mixed-methods approach, incorporating qualitative and quantitative data collection techniques such as in-depth interviews, focus group discussions (FGDs), key informant interviews and surveys from the flash flood-affected people residing in the five affected villages. The study observes that the flash flood caused severe damage to the agricultural sector, with almost every farmer in the affected area losing their single-season high-yielding *Boro* paddy, resulting in year-long food insecurity and unemployment. The government responded to the situation by providing year-long food and monetary aid through various social safety net programmes, although some accusations of nepotism and embezzlement were reported. The study found that the majority of the people were reliant on government aid for survival throughout the year. Moreover, some NGOs also played a crucial role by providing food and other support. The research suggests that transparency

*Corresponding author.

Growth, Poverty and Developmental Aspects of Agriculture, 123–142

doi:10.1108/978-1-83608-076-320241009

and accountability must be prioritized to ensure fairness, and positive lessons learned from the government's efforts during this flash flood might be helpful in reducing vulnerability and distributing relief more effectively.

Keywords: Flash flooding; food insecurity; disaster management; haor region; climate change; social safety net programme; Bangladesh

Introduction

Despite making significant economic progress in recent years, Bangladesh continues to face numerous challenges, including climate change disasters, unemployment and food insecurity. While the country has made strides in reducing poverty, with the poverty rate decreasing from 11.8% to 5.0% between 2010 and 2022, significant obstacles persist (World Bank, 2023). Bangladesh ranks seventh among nations impacted by climate-related disasters, with an estimated 51 million people experiencing food insecurity and 11 million suffering from acute hunger (WFP, 2024). According to the Labor Force Survey 2022, the total labour force in Bangladesh is 73.05 million, with the majority employed in agriculture (40.6%), industry (39%) and service (20.4%) sectors, while 4.7% remain unemployed (BBS, 2023; World Bank, 2022). However, climate disasters such as floods, cyclones, droughts and river erosion frequently affect people in Bangladesh, worsening food insecurity and unemployment, especially among lower income groups, due to income loss, property damage and loss of livestock and homesteads (Akter & Mallick, 2013; Khan, 2008; Zaman, 2019).

Throughout its history, Bangladesh has been formed by the gradual accumulation of flood deposits in the Bengal Delta. This geological process has rendered the region particularly prone to flooding, greatly influencing the country's landscape, culture and society (Hofer & Messerli, 2006). The phenomenon of climate change has triggered severe catastrophes in recent times, which have adversely affected the low-lying country of Bangladesh, particularly through the occurrence of recurrent and devastating floods. These floods are predicted to increase in frequency and magnitude, thereby compounding the challenges already faced by the country (Dastagir, 2015; Hossain et al., 2020). Floods have the potential to leave individuals exposed to a range of human suffering and financial setbacks, such as the tragic loss of lives, homes, food supplies, livestock, crops, educational opportunities, savings, sanitation and health. In turn, this can result in unemployment and illness for the victims, ultimately leading to profound misery and hardship for those impacted by the catastrophe (Hoq et al., 2021; Hossain et al., 2020; Rashid, 2000). Flash floods can cause more damage than other types due to their sudden onset, leaving little time for preparedness or coping (Kamal et al., 2018; Paul & Routray, 2010).

Located in the north-eastern region of Bangladesh and being interconnected to the Meghalaya and Assam states of India, the Sunamganj district is known for its unique wetland ecosystem, which is locally referred to as *haor*. This wetland ecosystem is characterized by heavy rainfall during the monsoon season, leading to floods that generally last from May or June and continue for approximately 7–8 months with a customary depth of 0–5 m. The inhabitants of this region are heavily dependent on the resources provided by these bowl-shaped *haors*, which are crucial for their livelihoods, culture and overall socio-economic needs. During the monsoon season, the *haors* are utilized for fishing purposes, and in the post-monsoon or dried period, they are used to cultivate high-yielding *Boro* paddy as a single crop, which is a critical source of food and income for the local populace (Chakraborty et al., 2021; Dey et al., 2021; Kamal et al., 2018; Zafar Sadique et al., 2017).

In April 2017, there was an unexpected and excessive amount of rainfall in the north-east India–Bangladesh region. This resulted in a sudden and severe flash flood in six interconnected districts, including Sunamganj. It should be noted that this flood was that it occurred 1 month earlier than usual, which had not happened in decades (Chakraborty et al., 2021; Zafar Sadique et al., 2017). This flash flood affected 4.67 million individuals from 1 million households, causing $450 million in damage to rice crops and adversely impacting livestock, schools and fisheries (Kamal et al., 2018; Zafar Sadique et al., 2017). The aftermath of the disaster resulted in significant challenges concerning unemployment and food insecurity, which posed a major threat to the affected people. In response to this predicament, the government, NGOs, and philanthropists extended their support to provide aid and alleviate the suffering of the disaster-affected population (Chakraborty et al., 2021; Dey et al., 2021; Roy & Blaschke, 2015; Zafar Sadique et al., 2017).

Several studies have been conducted to examine various aspects of the 2017 flash flood, including its causes, damages, impacts, vulnerabilities and resilience of the affected population. However, the literature has paid limited attention to the role of various stakeholders, such as the government, NGOs, voluntary organizations and philanthropists, in managing this catastrophe. Given the importance of preparing for future climate-related disasters in Bangladesh, it is imperative to investigate the responses of these stakeholders to such events. Thus, this study aims to scrutinize the approaches adopted by the government and NGOs in addressing critical issues, such as unemployment, food insecurity and other challenges faced by the people affected during the 2017 flash flood in the north-eastern region of Bangladesh. This inquiry will shed light on the effectiveness of the measures taken by these stakeholders and provide insights into the best practices that can be employed to manage similar disasters in the future.

Government and NGOs' Responses to the Natural Disasters in Bangladesh

The impact of natural disasters on the socio-economic and political history of Bangladesh has been significant, both prior to and following its independence. One of the devastating calamities was the Bhola cyclone, which struck East Pakistan (now Bangladesh) in 1970. The Pakistani military government's response to this natural disaster was deemed inhumane, leading to an estimated 500,000 fatalities. This calamity had a direct effect on the Bengali Nationalist Movement, ultimately paving the way for the Liberation War of 1971 and resulting in the independence of Bangladesh (Biswas & Daly, 2021; Hossain, 2018). Bangladesh has since faced many natural disasters, with some of the most severe ones including the 1988 flood, the catastrophic super cyclone of 1991, the devastating flood of 1998 and the tropical Sidr cyclone of 2007 (Islam & Walkerden, 2015; Karim, 1995; Zaman, 2019).

In response to the 1970 flood, Bangladesh established the Cyclone Preparedness Program (CPP) in 1972 (Hasan et al., 2019). Later, several devastating floods and cyclones followed, leading to the establishment of the Disaster Management Bureau (DMB) and the formulation of the Flood Action Plan (FAP), the Standing Orders on Disaster (SOD) 1997 (revised 2010) and the Comprehensive Disaster Management Programme (CDMP) in 2004 (Hasan et al., 2019; Uddin et al., 2021). Currently, Disaster Risk Reduction (DRR) and Emergency Response Management (ERM) programmes in Bangladesh are implemented according to the Disaster Management (DM) Act 2012, Disaster Management Policy 2015, SOD 2019, National Plan for Disaster Management (NPDM), Perspective Plan (2021, 2041), Eighth (8th) Five Year Plan and Bangladesh Delta Plan (BDP) 2,100 (GoB, 2020; Sabur, 2012).

The National Disaster Management Council (NDMC), led by the Prime Minister, is responsible for guiding the formulation of planning and policies for disaster management in Bangladesh. The Ministry of Disaster Management and Relief (MoDMR) serves as the secretariat to the NDMC and is responsible for implementing these policies. In order to address disaster risks in vulnerable communities, MoDMR follows the NPDM, which is updated every 5 years (GoB, 2020). This process involves the collaboration of relevant ministries and departments, field administration, local government institutions (LGIs) and civil-armed forces in disaster preparedness, emergency relief and rehabilitation activities (GoB, 2020; Hossain et al., 2020; Uddin et al., 2021). Moreover, NGOs, cultural voluntary organizations and philanthropists, aside from government aid, are crucial in aiding disaster-affected people in Bangladesh (Gros et al., 2023; Hossain et al., 2020; Islam & Walkerden, 2015; Paul & Routray, 2010).

The devastating disasters that have struck Bangladesh have had a profound impact on the food security and employment situation, particularly for the lower income groups residing in the affected areas (Akter & Mallick, 2013; Rahman et al., 2018; Zaman, 2019). In an effort to address this dire situation, social safety net programmes (SSNPs) have played a crucial role in Bangladesh (Hassan et al., 2015; Uddin et al., 2021). To mitigate the vulnerability and crisis of the people encountering disaster, the government extends its regular SSNPs, which include Vulnerable Group Feeding (VGF), Vulnerable Group Development (VGD), Open Market Sale (OMS) and other general relief activities funded through Block Allocation for Disaster Management (Coirolo et al., 2013; Hassan et al., 2015; Zaman, 2019). Additionally, other regular cash transfer SSNPs, such as Old Aged Allowance, Disable Allowance, Allowance for the Widow, Deserted and Destitute Women and Maternity Allowance for the Poor, have also had a positive impact during the recovery and reconstruction process (Asma et al., 2023; Coirolo et al., 2013; Islam & Walkerden, 2015; Khandker et al., 2011).

Some studies have been conducted to investigate the response of the government and NGOs during natural disasters in Bangladesh. Previous studies carried out by Ninno et al. (2001), Paul (2003), and Coirolo et al. (2013) have concluded that government aid plays a critical role during periods of vulnerability in catastrophic areas. Despite some limitations, field administration and LGIs generally implement disaster management and other development activities at the grassroots level (Hasan & Ara, 2022; Islam et al., 2017). The studies by Khan (1991) and Islam et al. (2017) revealed that LGIs collaborate with field administration to provide relief, sanitation services, health facilities, livelihood support and reconstruction activities during and after disasters in Bangladesh.

In 1988, Bangladesh encountered its most devastating flood after its independence, where 53 out of 64 districts, approximately 57% of the total land, were flooded (Office of the US Foreign Disaster Assistance, 1988; Rasid & Pramanik, 1993). The flood caused significant damage, leading to farmers and lower income groups falling into severe food insecurity, indebtedness, unemployment and migration to urban areas (Brammer, 1990; Office of the US Foreign Disaster Assistance, 1988; Shaw, 1989). The government and international aid donors, including the United States, France, the United Kingdom and the United Nations Development Programme (UNDP), provided technical and food assistance in response to this dire situation. As a result, 45% of Bangladesh's development budget for the 1988–1989 financial year was cut to provide ad hoc assistance to the affected population (Brammer, 1990). Shaw's (1989) observations shed light on the crucial role played by various entities in providing assistance to flood victims in urban areas during the 1988 floods.

Bangladesh was struck by a catastrophic cyclone in 1991, one of the most severe natural disasters of the century (Khalil, 1993). The cyclone featured wind speeds of up to 225 kph and generated up to a 6-metre tidal wave that devastated the southern coastal region, leaving over 145,000 people dead (Begum, 1993; Khalil, 1993). The disaster also caused over a billion dollars in damage to crops, livestock, homesteads, infrastructure and other property. The victims of the cyclone were afflicted with food insecurity and unemployment, leading to unprecedented hardship and suffering (Alam & Collins, 2010; Khalil, 1993). Rahman and Bennish (1993) conducted a study that partially shed light on the relief management process during this catastrophe. They reported that the government, international donors and NGOs provided support to vulnerable communities with the support of armed forces, field administration and LGIs, which was coordinated by an emergency relief coordination committee headed by the prime minister. Moreover, Begum (1993) observed the role of national and international NGOs during the relief distribution process, with the support of LGIs, field administration officials, armed forces and local politicians. Conversely, Alam and Collins (2010), in their study, reveal that providing critical support, such as food, clothing, shelter and other necessities, by relatively affluent neighbours and relatives during the initial crisis period was crucial.

In the year 1998, Bangladesh faced the most severe flood of the century that lasted for over 2 months, from mid-July to September, and affected more than two-thirds of the country (del Ninno & Lundberg, 2005; Rashid, 2000). The magnitude of this disaster led to a significant impact on the lives of approximately 127 million people. Despite its severity, the flood's impact on households was significantly reduced by the successful collaboration between the Bangladesh government, NGOs, donors, flood-affected households and private businesses operating in efficient markets (Ninno et al., 2001). However, Ahmed's (2001) study revealed some irregularities in the government's SSNPs. He discovered that political influence was present in the beneficiary selection and distribution process of the VGF, Food for Work and Test Relief programmes. While Ahmed acknowledged the role of NGOs in reducing the impact of the flood, he also found that their activities were limited to their members. Paul's, study (2003) showed that the government's relief management process during the emergency crisis was better than in previous disasters, and NGO support was crucial for the victims. Furthermore, Rashid (2000) explored the plight of lower income groups residing in urban areas and found that most of them lost their jobs, leading to food insecurity and health hazards.

In 2007, Cyclone Sidr wreaked havoc on the coastal region of southwestern Bangladesh, comparable in magnitude to the cyclone that struck the country in 1991. However, despite the severity of the disaster, the number of fatalities was significantly lower at 3,406 individuals (Haque & Jahan, 2016; Paul, 2009). This outcome can be attributed to the government's preparedness, as well as other significant factors such as the duration and nature of the storm, its surge, the time of landfall and the ecology and embankment of the coastal area (Paul, 2009). Nevertheless, the cyclone had a profound effect on various aspects of life, such as livelihood, food security, crops, homestead, sanitation, education and physical and mental health (Hawlader, 2011; Paul et al., 2011). Subsequently, the government and NGOs intervened during the recovery phases, and their efforts were lauded in various studies conducted by researchers, suggesting that their intervention had a direct impact on reducing casualties (Aase, 2020; Hawlader, 2011; Paul et al., 2011). However, some studies have revealed anomalies in the intervention process of NGOs and the government. For instance, Aase (2020) discovered that politicians and local elites played a significant role in selecting the beneficiaries for aid programmes. Similarly, Hawlader (2011) identified several cases of financial irregularities, such as not halting loan repayments during a crisis, giving preference to their own members, bribery during relief distribution and creating dependency on aid interventions by NGOs.

In response to the misery, unemployment and food insecurity caused by climate disasters in Bangladesh, the government, with the support of international donor agencies, has played a leading role in providing relief activities, sanitation services, health facilities, livelihood support and reconstruction activities during the recovery and post-recovery stages. While the government's efforts are commendable, studies(Aase, 2020; Ahmed, 2001; Alam & Collins, 2010; Asma et al., 2023; Hassan et al., 2015; Khalil, 1993) have revealed significant drawbacks in its disaster management for vulnerable communities. These include bribery and nepotism in the beneficiary selection process, misappropriation of relief funds, embezzlement of relief, underpayment and misuse of social power by the elite during and after disasters. On the other hand, NGOs have increasingly played a crucial role in disaster management in Bangladesh since the end of the 20th century. Studies by Gros et al. (2023), Islam et al. (2017), Islam and Walkerden (2015), and Khan (2008) have emphasized the essential role played by NGOs during and after disasters.

Methods (300)

This study aimed to examine the extent and the response of both government and NGOs to the misery, unemployment and food crises experienced by victims of the 2007 flash flood in the Sunamganj district. The research was conducted in five villages located in the two most severely impacted Upazilas (*Shulla* and *Tahirpur*) of Sunamganj district. To achieve the research objectives, a convergent parallel mixed approach was utilized, which enabled the collection and analysis of both qualitative and quantitative data concurrently, resulting in a more comprehensive understanding of the research problem (Creswell, 2012). Primary data were collected from household heads, community leaders and representatives of LGIs and NGOs in the study area from March to April 2018. Qualitative data were collected through in-depth interviews, focus group discussions (FGDs) and key informant interviews following the checklists developed from the 2017 flash flood-related news published in leading local and national newspapers and government and NGO reports.

To ensure that the data collected are representative, a sample of 25 household heads (five from each village) was selected using judgemental sampling for in-depth interviews, considering their age, income, number of family members, gender and extent of losses by the flash flood. Similarly, the participants of the five FGDs (two for women and three for men) were selected based on their age, income and the extent of damages. Twenty key informant interviews were conducted with the two Union Council (UP) chairmen, five UP members, five primary and secondary schoolteachers, three NGO representatives and five community leaders. All the interviews were recorded with the consent of the participants. In addition, a survey was conducted, and 100 household heads were selected using multistage random sampling techniques. Data were collected using a structured questionnaire.

For qualitative data analysis, we recorded all interviews in their native dialects and later translated them into English. After that, we coded and categorized the transcriptions and developed themes based on the study objectives. The quantitative data were analyzed using SPSS software, which utilized frequency distribution and graphs.

Study Findings

Unemployment, Food Insecurity and Miseries Caused by the 2017 Flash Flood

Socio-Economic Condition of the Respondents

The majority of the participants, who were primarily household heads, were engaged in farming (47%), followed by daily labour (19%) and fishing (15%). A small proportion of participants reported being involved in business (5%), while 8% were housewives and the remaining 6% engaged in other irregular activities. Furthermore, the study found that a significant proportion of the participants (67%) reported earning less than Tk. 5,000 (approximately 46 dollars) per month, while 33% of the respondents reported earning between Tk. 5,000 and Tk. 10,000 (46–91 dollars) on a monthly basis.

Affected Sectors and Their Extent

The flash flood has severely impacted most of the sectors connected to these families living in the area. Nearly 89% of the participants reported that their families were severely affected, while the remaining 11% claimed that the impact was comparatively less severe than others. The agricultural sector has suffered the most significant damage, with all the Boro crops in the area being destroyed by the flash flood. The study finds that 90% of the respondent's families lost their crops, and the remaining 10% of the respondents were not involved in agriculture. The homestead was the second-most-affected area, with 74% of the respondents' houses being impacted. The study also reveals that 57% of the respondents lost their source of daily earnings, while the percentage of people affected by the livestock was 52. Additionally, the study reports that 52% of the respondents have lost their fishing spaces due to the flash flood (Fig. 9.1).

Effect on Agriculture and Food Security

The livelihoods of the inhabitants residing in the *haor* area are heavily reliant on cultivating high-yielding single-season *Boro* paddy, which serves as their primary source of sustenance and income. The study has revealed that owing to the occurrence of flash floods, 100% of the farmers were unable to harvest the cultivated paddy from their fields, which had a severe impact on their overall food security. It has been found that a majority of farmers had borrowed money from NGOs and local moneylenders to cover the expenses of purchasing seeds,

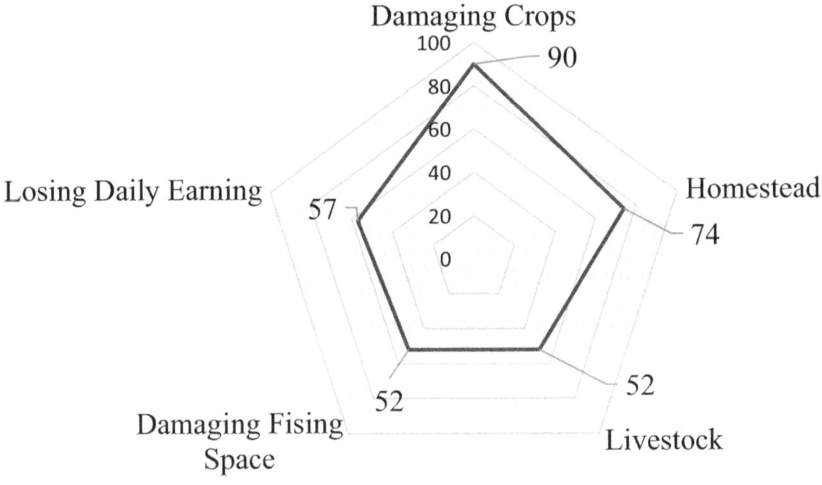

Fig. 9.1. Affected Areas Caused by the 2017 Flash Flood (Source: Survey Data). *Source:* Field survey.

fertilizers, pesticides and other necessary items during the cultivation process. These loans usually are repaid after harvesting their crops. However, due to unforeseen circumstances this year, farmers were unable to pay back their loans. The lenders charged high-interest rates, which further exacerbated the financial burden on the farmers. One of the farmers stated during his interview:

> This year, 'dhal' (locally recognised as the flash flood) came a month ahead of usual. This meant that Paddy did not have enough time to mature or ripen properly. Consequently, nobody was able to harvest their crops and bring them home. You know this is the only source of our income and food, which has been entirely lost. In addition, we are burdened with repaying the loans that we took from NGOs and moneylenders to cover the expenses of buying seeds, fertilisers, pesticides, and labour costs. This has left us in a difficult position that is unimaginable for others.

The haor regions, where fishing is a crucial source of livelihood for the local population, were severely impacted by the flash flood in 2017. The resulting damage to the area led to reports of dead fish, frogs and birds, giving rise to initial rumours that open pits in Meghalaya were contaminating the area with uranium (Ahmed & Roy, 2017). As a result, the local administration imposed a ban on fishing in the affected area, which has had significant negative implications for the people's livelihoods and their ability to recover from food insecurity. The ban hindered recovery efforts in the area, exacerbating the already dire food insecurity situation. The participants of an FGD argued:

> In previous years, whenever our land was flooded, we used to catch fish from the flooded areas to help us recover from the flood damages. However, this year, we faced serious trouble when the administration banned fishing in the haor. We have never experienced this before, as we have never seen so many fish die immediately after a flood. Firstly, we lost our primary food source, paddy, and subsequently, the ban on fishing closed another vital source of food and income, worsening our immediate misery. However, a few weeks later, the authorities lifted the ban. By then, we had already started receiving relief and other assistance from the government and NGOs.

Effects on Livestock

Livestock was identified as one of the sectors that was significantly impacted by the flash flood. The survey results revealed that 74% of respondents were compelled to sell their livestock due to coping with the immediate financial crisis and shortage of fodder after the flood. Furthermore, a notable proportion of households (20%) reported that their livestock had perished due to starvation, while a smaller proportion (6%) indicated that their livestock had been swept away and lost in the flood. Qualitative data from the study suggest that paddy straw is the primary source of fodder for their cattle; however, it was unavailable due to the flooding of paddy fields. As a result, the cost of fodder has increased by a significant margin, ranging from three to four times higher than the typical period, forcing farmers to sell their livestock. Moreover, the immediate economic and food crisis compelled flood victims to sell their livestock at a lower price than usual, exacerbating the financial strain caused by the flood. One of the survivals of flashflood shared:

> On the first night, after breaking our embankment, my house and surroundings got flooded. Unfortunately, my chickens and a calf got washed away and lost in the flood. I wasn't able to make any preparations prior to the flood as I had no idea it would come that night. Eventually, due to the unavailability of food for them and our own food arrangements, I was forced to sell my other cow and goat at half price compared to usual. As we couldn't harvest paddy, the scarcity of paddy straw was higher than ever. So we had to sell these animals for our immediate survival.

Effects on Daily Earning

In 2017, an untimely flash flood had a severe impact on the socio-economic fabric of the haor area, resulting in a significant loss of livelihoods and income sources for its inhabitants. The deluge inflicted extensive damage on the primary sources of employment in the region, including farming, fishing and daily wage labour, leading to a multitude of job losses and reduced income opportunities. According to survey data, 57% of the respondents were affected by this loss. Furthermore, the damage caused by the flood had far-reaching consequences on the indirectly dependent sectors, such as local businesses, which rely heavily on the community's economic well-being. As a result, the flood's adverse impact on the region's overall prosperity was aggravated, leading to an extended period of economic hardship for its people. During the in-depth interview, one of the grocery owners shared:

> This early flood had a direct or indirect adverse effect on everyone's livelihood in this area. As you are aware, most of the jobs in this region are related to the 'haor'. During the dry season, people farm on the haor land, while during the monsoon season, many shift to fishing. Moreover, daily labourers primarily work on the farming land of this haor. Other minor jobs, such as carpentry, painting, barbering, or small businesses like mine, are indirectly dependent on the income of these people. If they fall into economic turmoil, our businesses will suffer as well, as I am currently

> experiencing. Since the beginning of this disaster, I have been struggling with my family and relying on government assistance to make ends meet.

Effects on Homestead and Education

The survey revealed that a substantial percentage of respondents reported damage to their homesteads. However, the qualitative analysis suggested that the impact of the flash flood on the homesteads was not as catastrophic as other natural disasters, such as cyclones or natural floods. This finding could be attributed to the unique characteristics of the flash flood, which overflowed the wetland and damaged the lower part of the homesteads, thus affecting a significant numbers of the latrines, tube well and gardening situated in the lower part of the homestead more than the houses. Moreover, the analysis indicated that the houses in Tahirpur Upazila were more vulnerable to damage than those in Shulla Upazila. It was found that the damage incurred by the embankment of the study villages resulted in severe tides that damaged the houses in the lower part of the villages.

The flash floods had a significant impact on the educational institutions of the region. As a result of the flood, both primary and secondary schools had to close for a week or more. The flood inflicted physical damage to the infrastructure of numerous schools, particularly those situated in the Tahirpur Upazila. Despite the reopening of schools, student attendance remained low, and the affected schools required almost a year to return to normalcy. The flash flood had a significant impact on the communication system, causing many students to be absent from school as they had to assist their parents in economic activities to cope with the crisis. According to key informant interviews, a considerable number of female students migrated to urban areas in search of employment opportunities to support their families. During an interview, an assistant head teacher at a secondary school in the Shulla Upazila acknowledged this migration trend and said:

> We were informed that prior to the final examination, some guardians of our female students requested permission for their wards to take the exam. The reason behind this request was that these students were unable to attend classes as they had to work in garment factories in Dhaka. They did this to support their families in recovering from the aftermath of the flood. Additionally, many migrant students haven't returned to school since and might be continuing their jobs there.

Government, NGOs and Other Responses

The flash flood victims of the affected area received assistance from the government, NGOs and philanthropic organizations after a delay of 1 to 2 weeks.

However, by the time aid reached the victims, they had already undergone extreme levels of misery as their neighbours were unable to provide any assistance due to the widespread impact of the disaster. As a consequence, the victims were compelled to sell their valuable items, such as livestock and jewellery, at half-price to afford food and other essentials. The sellers were found to be struggling to manage buyers, leading to a decline in prices. The survey findings and in-depth interviews revealed that most families were unable to afford three meals a day before receiving assistance. During the FGDs, it was found that women participants prioritized feeding children and elderly family members first, followed by male adults and women last. In most cases, women experienced starvation more than male family members. Additionally, the study participants reported an increase in theft in every area, including the stealing of fishing nets and other essentials. This had an adverse effect on coping with the early stage as the inhabitants of the haor area usually depend on fishing after flooding every year.

In response to the flash floods that brought immense suffering to its victims, the government promptly administered relief through various programmes such as Gratuitous Relief (GR), VGF and OMS. The GR programme provides 20 kg of rice per individual for around a year to flash flood victims. The VGF programme provided 30 kg of rice and Tk. 500 (equivalent to about $4) per month for selected households. The OMS programme was operational throughout the year, providing rice at Tk. 15 per kg until September 2017, after which the price was refixed at Tk. 30 per kg. In addition, the government distributed woollen cloth during winter and provided farmers with 5 kg of seeds, 30 kg of fertilizer and Tk. 1,000 (equivalent to approximately $8) at the beginning of the *Boro* farming season. Furthermore, the government implemented special programmes in the area, such as VGD and Money for Work *Kajer Binimoye Taka* (KABITA), in addition to its regular SSNP, e.g. Allowances for Widows and Old People.

With the assistance of LGIs, the field administration executed all government-supporting programmes in the rural affected areas. Various committees coordinated their efforts at each Upazila and Union level, with the Upazila Executive Officer (UNO) and Union Council (UP) Chairman serving as chairs, respectively. At the Union Parishad (UP) level, Ward Committees (WC) were formed at each Ward led by UP members were established, with other members comprising female UP members, prominent local figures and community leaders. The aforementioned committees played a crucial role in the beneficiary selection and relief distribution process in affected villages.

Some NGOs and voluntary organizations played a critical role in providing support to those affected by the floods in the study area. Brac, which is the world's largest NGO and started its operations in the study area in 1972, was particularly instrumental in providing various forms of assistance to those affected by the 2017 flash floods. Other organizations, like World Vision and Care Bangladesh, also extended their support to the flood victims in Tahirpur Upazila.

The study's findings indicate that government assistance played a pivotal role in the survival of flood victims. Survey data revealed that 86% of respondents received support from the government during the crisis. Notably, 69% of these individuals relied solely on government assistance to survive after the flash flood.

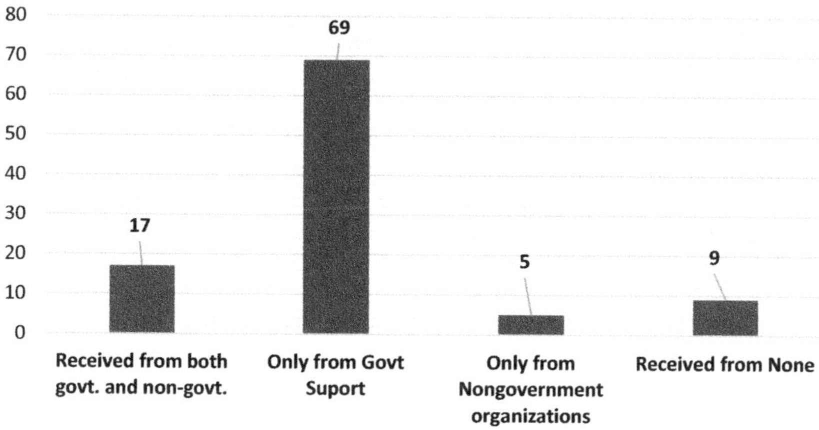

Fig. 9.2. Sources of Different Sorts of Assistances. *Source:* Field survey.

In comparison, only 22% of respondents received aid from NGOs, with a mere 5% indicating that they depended entirely on NGO assistance. It is noteworthy that 9% of the respondents did not receive any aid from any source (Fig. 9.2).

The survey findings report a significant proportion of the government beneficiaries, precisely 85%, acknowledged receiving a monthly ration of 30 kg of rice and Tk. 500 throughout the year. Moreover, 28% of the respondents opted to procure rice from the OMS programme. The collected data from FGD and in-depth interviews highlighted the pivotal role played by the government-supported year-long programmes in aiding the survival of individuals affected by the food crisis caused by flash floods. One of the survivors of Shulla Upazila shared during the in-depth interview:

> After a week of the flash flood, the government finally started providing relief. While some NGOs and charitable organisations also offered assistance to a few people, unfortunately, the majority of the affected individuals did not receive their aid, and their assistance was provided only once or twice. The government's aid was the most valuable for us, and it continued throughout the year. Additionally, every vulnerable family in this area received some form of government assistance, which helped them to survive. Although the amount of food and money provided by the government wasn't substantial, it was still better than having nothing to survive on and potentially starving to death.

The process of selecting beneficiaries proved to be a challenging task due to a scarcity of VGF and VGD cards in relation to the number of affected families.

The Shulla UP of Shulla Upazila, for instance, received 2,820 cards from the government, whereas the total number of affected families stood at 4,800. This insufficiency presented an unprecedented challenge for the local government officials and field administration to provide assistance to all the affected families. Consequently, the officials excluded the families who had already received any form of support from SSNP and those who were receiving aid from any NGO or were financially better off than others. The UP chairman of *Shulla*, who led the beneficiary selection process of the government assistance in his union council, stated these challenges:

> We have received 2820 VGF and VGD cards from the government. Each card includes 30 kg of rice and Tk. 500 to provide to the flood victims every month. However, the total number of families affected by the flash flood in our UP was approximately 4800. This posed a significant challenge to us in ensuring the proper distribution of government aid. To address this issue, we implemented certain strategies, such as excluding families that were already included in another or regular SSNP or receiving support from an NGO. We also did not include families whose members were employed in a government or other job and could afford the crisis with their earnings. We knew all the families personally, and we were aware of how much the flood had affected their families and their economic condition, which helped us select the right beneficiaries.

Some accusations of nepotism and the influence of local elites in beneficiary selection and misappropriation in the relief distribution process were reported in the study. Qualitative findings of the study indicated that most of the accusations were related to the beneficiary selection process and rice weighing less than due. However, the UP chairman denied that accusation and shared:

> The beneficiary selection process was closely monitored by the UNO office and members of the ward committees, which consisted of UP members and community leaders. This ensured that there was little room for nepotism. Additionally, digital machines were used to weigh the rice, preventing any possibility of underweight relief. However, some beneficiaries were accustomed to selling their rice in the local market, where buyers could potentially cheat them by offering less money and accusing the officials.

The findings of the study reported that the aid provided by NGOs during times of crisis was deemed inadequate by the participants. Notably, most NGOs prioritized their microcredit borrowers over other affected individuals. However, Brac was an exception in Shulla, as it provided assistance beyond its microcredit borrowers. Unfortunately, the assistance from most NGOs was limited to their members and was provided only once or twice a year. In contrast, Brac provided unique and unconventional aid forms such as *Shikkha Tori*, boat services for students to transport them to their schools and cattle food, which were crucial for the affected individuals to return to everyday lives.

The findings of the study indicate that the aid provided by voluntary or charitable organizations played a crucial role in mitigating the initial stage of the crisis faced by the participants. However, the process lacked proper management and coordination with the local administration. The relief efforts were executed without any prior needs assessment, which resulted in an inefficient allocation of resources. Additionally, the relief distribution was often influenced by the gatekeeper, who directed the aid workers to distribute the relief to their acquaintances, such as family members, relatives and neighbours. Therefore, the study participants pointed out that the gatekeeper's connections were more benefited than other vulnerable victims. One of the key informant shared:

> Philanthropic and charitable organizations often distribute their relief without conducting a prior study of the vulnerability of disaster-affected individuals. Instead, they rely on the guidance of someone who may have influenced their decision in selecting beneficiaries, such as family members, relatives, or members of the community affiliated with the guide. As a result, some people receive assistance that they need less than other severely affected victims. To make the relief efforts more effective, it is important for these organizations to communicate with local authorities before selecting their relief area and beneficiaries.

Discussion

The research findings reveal that the flash flood that occurred in Bangladesh in 2017 bears several resemblances to previous natural disasters that have struck the country, particularly with respect to the suffering and unemployment that ensued. However, this calamity stands out owing to its unprecedented devastation of crops, leading to a situation where all farmers lost their single crops for the year. Consequently, there has been a substantial increase in food insecurity for the entire population, which is a unique outcome when compared to other disasters where primarily the lower income groups have experienced severe food insecurity, as noted in previous studies by Akter and Mallick (2013), Hawlader (2011), and Paul (2003). Furthermore, the findings provide evidence for Begum's (1993) assertion that women are more vulnerable than men in disasters. Conversely, the study indicated that the extent of direct adverse effects on properties, such as damage to houses and loss of livestock and other valuables, was comparatively less than other disasters.

The study found that the assistance provided by the government during the 2017 flash flood played the most vital role in supporting flood victims who were grappling with food insecurity and unemployment. However, unlike the 1998 flood (Brammer, 1990; Shaw, 1989) and the 1991 super cyclone (Rahman & Bennish, 1993), there was no direct involvement in relief assistance from international donor agencies or countries during the 2017 flash flood. The study highlights the crucial and collaborative role of the field administration and LGIs in the disaster management process, as discussed in earlier studies by Khan (1991) and Islam et al. (2017). Furthermore, the study reported fewer allegations of bribery, nepotism and political influences against government officials and LGI representatives during the 2017 flash flood than in previous disasters, such as those reported by Aase (2020) and Ahmed (2001). The study also sheds light on NGOs' vital roles during the flash flood. However, the study found that the magnitude of their involvement was not as significant as in previous disasters such as the 1998 flood (Ninno et al., 2001) and the Sidr cyclone of 2007 (Paul, 2009). Notably, the study found no allegations of NGO corruption related to flash flood victims, unlike in the Sidr cyclone of 2007, as reported by Hawlader (2011).

The findings of the study exhibit certain similarities with the experiences of Shaw (1989) and Alam and Collins (2010), which indicate that voluntary organizations and philanthropists are instrumental in helping disaster victims survive previous catastrophes. Furthermore, previous studies conducted by Alam and Collins (2010) and Shaw (1989) highlighted the significance of social capital or support from neighbours in facilitating disaster recovery in Bangladesh. However, the findings of this study contradict their observation, as it reveals that the victims of the 2007 flash flood received minimal assistance from their neighbours and surroundings due to their high vulnerability.

The research findings indicate that the management of the flash flood catastrophe that occurred in Bangladesh in 2007 was efficiently executed compared to previous disasters. Particularly, the nature of the assistance run by the government for this disaster was unique, as it was prolonged for over a year and had to provide assistance to a vast number of flood victims who were severely affected by food insecurity and economic losses. However, this study also revealed a coordination gap between the government and NGOs in the relief distribution process, which needs to be addressed for future disaster management. Prior coordination can also help to ensure that resources are used more effectively and efficiently, thereby maximizing the impact of disaster management efforts. Furthermore, the government should increase its coverage of SSNPs, as the study found that government programmes played a crucial role in saving disaster survivors during a 2017 flash flood.

Conclusion

Bangladesh has historically been a country that has faced severe natural disasters, with its people losing their lives, properties, livelihoods, infrastructure, income and livestock, leaving them extremely vulnerable. In response to this vulnerability, the government, along with donor agencies, NGOs and charitable organizations, has extended its assistance to disaster victims. Over the years, the government has adopted policies and acts aimed at improving disaster management, with special emphasis on coordinating various ministries, departments, agencies, LGIs, NGOs and other relevant forces. As a result, during previous natural disasters, the government played a crucial role in reducing the misery and vulnerability of the disaster-affected people by providing assistance. The flash flood of 2017, caused by excessive rainfall and onrush of water, flooded the north-eastern six districts, including the Sunamganj district of Bangladesh, causing unprecedented food and unemployment vulnerability to the *haor*-based population. Almost all farmers in this area lost their single-season high-yielding *Boro* paddy, which caused them year-long food insecurity and unemployment. The victims received no support from their neighbours due to the extent of their vulnerability. The government had to provide year-long food and money assistance programmes through the VGF, VGD, OMS and other regular SSNPs, which were unique in duration. While government assistance was not substantial enough, however, the majority of the affected population's survival was solely dependent on government assistance throughout the year.

The study recommends greater coordination among NGOs and voluntary organizations in disaster response efforts, emphasizing the need for transparency and fairness in the beneficiary selection and relief distribution process. Ultimately, the study suggests that the positive lessons learnt from the government's assistance management during the 2017 flash flood may be applied in future disaster response efforts to reduce the vulnerability of affected populations and ensure more effective and efficient relief distribution.

References

Aase, M. (2020). Listing for change? Exploring the politics of relief lists in Bangladesh after Cyclone Sidr. *Disasters*, *44*(4), 666–686.

Ahmed, I. (2001). Governance and flood: Critical reflections on the 1998 deluge. *Futures*, *33*. www.elsevier.com/locate/futures

Ahmed, R., & Roy, P. (2017, April 21). Uranium behind deaths in haors? *The Daily Star*. https://www.thedailystar.net/frontpage/uranium-behind-deaths-haors-1394068

Akter, S., & Mallick, B. (2013). The poverty-vulnerability-resilience nexus: Evidence from Bangladesh. *Ecological Economics*, *96*, 114–124.

Alam, E., & Collins, A. E. (2010). Cyclone disaster vulnerability and response experiences in coastal Bangladesh. *Disasters*, *34*(4), 931–954.

Asma, K. M., Misu, F., & Islam, M. A. (2023). The impact of social safety net programs on poverty, food security and standard of living in Bangladesh. *International Journal of Community and Social Development*, *5*(1), 71–96.

Bangladesh Bureau of Statistics (BBS). (2023). *Labour force survey 2022: Bangladesh*. https://bbs.portal.gov.bd/sites/default/files/files/bbs.portal.gov.bd/page/b343a8b4956b_45ca_872f_4cf9b2f1a6e0/2023-10-25-07-38-4304abd7a3f3d8799fbcb59ff91007b1.pdf

Begum, R. (1993). Women in environmental disasters: The 1991 cyclone in Bangladesh. *Gender and Development*, *1*(1), 34–39. https://doi.org/10.1080/09682869308519953

Biswas, S., & Daly, P. (2021). 'Cyclone not above politics': East Pakistan, disaster politics, and the 1970 Bhola Cyclone. *Modern Asian Studies*, *55*(4), 1382–1410.

Brammer, H. (1990). Floods in Bangladesh: Geographical background to the 1987 and 1988 floods. *The Geographical Journal*, *156*(1), 12.

Chakraborty, D., Mondal, K. P., Islam, S. T., & Roy, J. (2021). 2017 flash flood in Bangladesh: Lessons learnt. In *Disaster resilience and sustainability: Adaptation for sustainable development* (pp. 591–610). Elsevier.

Coirolo, C., Commins, S., Haque, I., & Pierce, G. (2013). Climate change and social protection in Bangladesh: Are existing programmes able to address the impacts of climate change? *Development Policy Review*, *31*(Suppl. 2), o74–o90.

Creswell, J. W. (2012). *Educational research* (4th ed.). Pearson.

Dastagir, M. R. (2015). Modeling recent climate change induced extreme events in Bangladesh: A review. *Weather and Climate Extremes*, *7*, 49–60.

del Ninno, C., & Lundberg, M. (2005). Treading water. The long-term impact of the 1998 flood on nutrition in Bangladesh. *Economics and Human Biology*, *3*(1), 67–96.

Dey, N. C., Parvez, M., & Islam, M. R. (2021). A study on the impact of the 2017 early monsoon flash flood: Potential measures to safeguard livelihoods from extreme climate events in the haor area of Bangladesh. *International Journal of Disaster Risk Reduction*, *59*(March), 102247.

Government of the People's Republic of Bangladesh (GoB). (2020). *National plan for disaster management (2021–2025): Action for disaster risk management towards resilient nation*. https://modmr.portal.gov.bd/sites/default/files/files/modmr.portal.gov.bd/page/a7c2b9e1_6c9d_4ecf_bb53_ec74653e6d05/NPDM%202021-2025%20Draft.pdf

Gros, C., Pronti, A., Sheikh, K., Hassan, A., & Shahjahan, M. (2023). Effects of anticipatory humanitarian cash assistance to households forecasted to experience extreme flooding: Evidence from Bangladesh. *Hydrology Research*, *54*(11), 1315–1328.

Haque, A., & Jahan, S. (2016). Regional impact of Cyclone Sidr in Bangladesh: A multi-sector analysis. *International Journal of Disaster Risk Science*, *7*(3), 312–327.

Hasan, M. M., & Ara, J. G. (2022). Political economy of local level budgeting in Bangladesh: A critical analysis. *Space and Culture, India*, *10*(2), 16–27.

Hasan, Md. R., Nasreen, M., & Chowdhury, Md. A. (2019). Gender-inclusive disaster management policy in Bangladesh: A content analysis of national and international regulatory frameworks. *International Journal of Disaster Risk Reduction*, *41*, 101324.

Hassan, R., Islam, M., Saifullah, A., & Islam, M. (2015). Effectiveness of social safety net programs on community resilience to hazard vulnerable population in Bangladesh. *Journal of Environmental Science and Natural Resources*, *6*(1), 123–129.

Hawlader, B. C. (2011). Socio-economic impact of Cyclone Sidr on rural livelihoods – A study of select villages in Barguna District, Bangladesh. *Asia-Pacific Journal of Rural Development*, (1). https://doi.org/10.1177/1018529120110106

Hofer, T., & Messerli, B. (2006). *Floods in Bangladesh history, dynamics and rethinking the role of the Himalayas floods in Bangladesh: History, dynamics and rethinking the role of the Himalayas*. United Nations University Press.

Hoq, M. S., Raha, S. K., & Hossain, M. I. (2021). Livelihood vulnerability to flood hazard: Understanding from the flood-prone haor ecosystem of Bangladesh. *Environmental Management*, *67*(3), 532–552.

Hossain, N. (2018). The 1970 Bhola cyclone, nationalist politics, and the subsistence crisis contract in Bangladesh. *Disasters*, *42*(1), 187–203.

Hossain, B., Sohel, M. S., & Ryakitimbo, C. M. (2020). Climate change induced extreme flood disaster in Bangladesh: Implications on people's livelihoods in the Char Village and their coping mechanisms. *Progress in Disaster Science*, *6*.

Islam, R., & Walkerden, G. (2015). How do links between households and NGOs promote disaster resilience and recovery?: A case study of linking social networks on the Bangladeshi coast. *Natural Hazards*, *78*(3), 1707–1727.

Islam, R., Walkerden, G., & Amati, M. (2017). Households' experience of local government during recovery from cyclones in coastal Bangladesh: Resilience, equity, and corruption. *Natural Hazards*, *85*(1), 361–378.

Kamal, A. S. M. M., Shamsudduha, M., Ahmed, B., Hassan, S. M. K., Islam, M. S., Kelman, I., & Fordham, M. (2018). Resilience to flash floods in wetland communities of northeastern Bangladesh. *International Journal of Disaster Risk Reduction*, *31*(June), 478–488.

Karim, N. (1995). Disasters in Bangladesh. *Natural Hazards*, *11*(3), 247–258.

Khalil, G. M. (1993). The catastrophic cyclone of April 1991: Its impact on the economy of Bangladesh. *Natural Hazards*, *8*(3), 263–281.

Khan, M. M. I. (1991). The impact of local elites on disaster preparedness planning: The location of flood shelters in northern Bangladesh. *Disasters*, *15*(4), 340–354.

Khan, M. S. A. (2008). Disaster preparedness for sustainable development in Bangladesh. *Disaster Prevention and Management: An International Journal*, *17*(5), 662–671. https://doi.org/10.1108/09653560810918667

Khandker, S. R., Abdul, M., Hussain, K., The, A. S., & Bank, W. (2011). *Can social safety nets alleviate seasonal deprivation? Evidence from northwest Bangladesh*. http://econ.worldbank.org

Ninno, C. D., Dorosh, P. A., Smith, L. C., & Roy, D. K. (2001). *The 1998 floods in Bangladesh: Disaster impacts, household coping strategies, and response*. International Food Policy Research Institute. https://ebrary.ifpri.org/digital/collection/p15738coll2/id/125308

Office of the US Foreign Disaster Assistance. (1988). *Disaster case report: Bangladesh floods FY 1988*. https://pdf.usaid.gov/pdf_docs/PBAAH117.pdf

Paul, B. K. (2003). Relief assistance to 1998 flood victims: A comparison of the performance of the government and NGOs. *Geographical Journal*, *169*(1), 75–89.

Paul, B. K. (2009). Why relatively fewer people died? The case of Bangladesh's Cyclone Sidr. *Natural Hazards*, *50*(2), 289–304.

Paul, B. K., Rahman, M. K., & Rakshit, B. C. (2011). Post-Cyclone Sidr illness patterns in coastal Bangladesh: An empirical study. *Natural Hazards*, *56*(3), 841–852.

Paul, S. K., & Routray, J. K. (2010). Flood proneness and coping strategies: The experiences of two villages in Bangladesh. *Disasters*, *34*(2), 489–508.

Rahman, M. O., & Bennish, M. (1993). Health related response to natural disasters: The case of the Bangladesh cyclone of 1991. *Social Science & Medicine*, *36*(7), 903–914.

Rahman, H. M. T., Mia, M. E., Ford, J. D., Robinson, B. E., & Hickey, G. M. (2018). Livelihood exposure to climatic stresses in the north-eastern floodplains of Bangladesh. *Land Use Policy*, *79*, 199–214.

Rashid, S. F. (2000). The urban poor in Dhaka city: Their struggles and coping strategies during the floods of 1998. *Disasters*, *24*(3), 240–253.

Rasid, H., & Pramanik, M. A. H. (1993). Areal extent of the 1988 flood in Bangladesh: How much did the satellite imagery show? *Natural Hazards*, *8*.

Roy, D. C., & Blaschke, T. (2015). Spatial vulnerability assessment of floods in the coastal regions of Bangladesh. *Geomatics, Natural Hazards and Risk*, *6*(1), 21–44.

Sabur, A. K. M. A. (2012). Disaster management system in Bangladesh: An overview. *India Quarterly*, *68*(1), 29–47.

Shaw, R. (1989). Living with floods in Bangladesh. *Anthropology Today*, *5*(1), 11.

Uddin, M. S., Haque, C. E., & Khan, M. N. (2021). Good governance and local level policy implementation for disaster-risk-reduction: Actual, perceptual and contested perspectives in coastal communities in Bangladesh. *Disaster Prevention and Management: An International Journal*, *30*(2), 94–111.

World Bank. (2022). *Unemployment, total (% of total labor force) (modeled ILO estimate) – Bangladesh*. https://data.worldbank.org/indicator/SL.UEM.TOTL.ZS?locations=BD

World Bank. (2023, October 4). *Bangladesh overview: Development news, research, data*. https://www.worldbank.org/en/country/bangladesh/overview

World Food Programme (WFP). (2024). *Bangladesh*. World Food Programme. https://www.wfp.org/countries/bangladesh

Zafar Sadique, M., Research Associate, S., & Bari, E. (2017). *Flood 2017: Assessing damage and post-flood management study team*. https://cpd.org.bd/resources/2017/10/Flood-2017-Assessing-Damage-and-Post-flood-Management.pdf

Zaman, M. Q. (2019). Vulnerability, disaster, and survival in Bangladesh. In *The angry earth* (pp. 162–177). Routledge. https://doi.org/10.4324/9781315298917-21

Chapter 10

Understanding the Problem of Food Security of the G-20 Countries From the Nexus Between Food Price and Food Production

Sudipta Jha and Biswajit Nayak

Vidyasagar University, India

Abstract

The problem of food security depends highly on food availability along with access, utilization and stability. In this study, we would like to focus on the food availability criterion. With the rise in food prices, production of food is expected to increase in the domestic as well as in the international market. Therefore, the G-20 countries collectively, which produce up to 80% of the world's cereals and also have the largest percentage share of world agricultural exports, ought to benefit from the rise in agricultural prices. However, a significant number of countries in this group are net importers in terms of agricultural inputs. Also, the rise in the prices of non-renewable resources raises the demand and cost of biofuels. Commodity trading and the weakening of the exchange rate also affect farm prices. A rise in farm prices from those accounts may not be supportive of to rise in food production and food security. Hence, it will be interesting to analyze the effect of rising food prices on the problem of food security of the G-20 countries from the perspective of food production. In this study, we have taken rice as a representative food item and found that rice production is significantly responsive to the domestic relative price of rice. However, there is no price incentive operative for rice productivity. Rather, we find evidence of a negative relationship between rice productivity and domestic rice prices in relation to the international prices.

Keywords: Food security; G-20; food price; food production; correlation; panel unit root; panel cointegration; panel causality

Growth, Poverty and Developmental Aspects of Agriculture, 143–162

Published under exclusive licence by Emerald Publishing Limited
doi:10.1108/978-1-83608-076-320241010

Introduction

The problem of food security depends very much on food availability along with access, utilization and stability. In this study, we would like to focus on the food availability criteria by focusing on food production. It is predicted that world agricultural production will need to rise by 70% by 2050, and specifically by nearly 100% in the developing countries, to feed a world population which is expected to reach over 9 billion by that time (G-20, 2011). The nature of agriculture is such that every nation is not capable of producing every product of the same quality and with the same efficiency. Therefore, food security in the modern era means not only a rise in per capita production but also a successful engagement in the international trade both as an exporter and an importer. Today the main food exporter countries are the major industrial countries like such as the United States, Germany, and France and a few developing countries like China, Brazil, and Argentina. The G-20 countries collectively produce up to 80% of the world's cereals and also have the largest percentage share of world agricultural exports. Therefore, understating the production prospects of these countries is crucial in analyzing the food security problem.

Barring the most recent uninvited events like COVID, the Russia–Ukraine war and the resultant food supply disruptions in the recent past, the world as a whole witnessed a steep rise in food prices in early 2000. The phase of price surge continued for the first 10 years of the new millennium, though from 2005, the terms of trade started to move against agriculture again. Even though agricultural prices now have come down considerably relative to 2011, in real terms, they are still 40% above their 2000 lows (Baffes & Haniotis, 2016). Definitely, this increase in prices raises concern over food security worldwide (see Heltberg et al., 2012; OECD, 2008; World Bank, 2008). However, with the rise in food prices, both production and exports of food are also expected to increase in the domestic as well as in the international market. With the improvement of terms of trade in agriculture, not only food availability but also investments and thus productivity of the agricultural sector are expected to move upward. Therefore, from this perspective, a remunerative farm price seems to be the precondition of food security. Many economists like Haile et al. (2014) argued that price volatility, not the price rise, is responsible for output disruptions. G-20 (2011) raised their concern over the fact that in developing countries, small-scale agricultural farmers are simultaneously the major producers and the most food-insecure people. Hence, improving their revenue and output will solve both the problem of food security on a large scale and supply disruptions of food for domestic and international markets. Now, the basic requirement for this idea is to provide a remunerative price to these farmers. Headey (2018) found empirical support for this claim.

However, what drives the world agricultural price escalation is a cause of concern. Without a commensurate rise in output, a rise in food prices at the national level, due to a rise in food demand in the urban sector, export demand or reduction in food import whatever the source, has a severe negative effect on wage workers, including those associated with agriculture.[1] To mitigate this kind of problem, Gouel (2012) argued for

[1]This is because the agricultural producer has to pay for all other food items except one which he/she produce.

government intervention if price fluctuation arises from forecasting errors. To some economists like Wright (2012), this global spike is related to reduction in stocks, induced by strong biofuel production. However, this claim is debatable, and there exists considerable literature on this debate (see Baffes & Haniotis, 2010; Brümmer et al., 2013; Clapp & Cohen, 2009; Gilbert, 2010; Headey & Fan, 2008; Tadesse et al., 2014). Krugman (2008) and Wolf (2008) pointed increase in population and related food demand rise in developing countries like China and India as a contributing factor in food price hikes. 'Financialization' of the commodities, depreciation of the US dollar and interest rates are considered as contributing factors in this context (see Abbott et al., 2008; Baffes & Haniotis, 2016; Elliott, 2008; OECD and FAO, 2008).

Another important thing is to note that the commodity producers are directly linked to the global commodity prices, not the consumers. Studies like Dewbre et al. (2008) and García-Germán et al. (2015) found a distinction between high international prices for food and the consumer costs of food. Food consumers do not buy raw food commodities at international prices.

However, at the macro level, agricultural production as a whole, crop diversification and cropping patterns do not change or respond instantaneously to price incentives. Similar to the industrial sector, the agricultural sector does not run with excess capacity. Therefore, it cannot alter its overall output in the short term in a significant manner with the fluctuations in prices. Traditional practices, concern over assured income, risk factors, technology, farmers' attitudes all play significant roles in this regard. Short-term production fluctuation is more of a production function's technical issue than an economic one in the agricultural sector. Economic incentives can impact agricultural production or more specifically, to our interest, food production, on a sustained basis only in the long term. Apart from technological innovation, we can expect an increase in food production in the long term with the increase in the relative prices of food. However, as reported by the Food and Agriculture Organization (FAO) (2022), farmers will fail to respond to the increase in food prices if the prices of agricultural inputs including food, fertilizers, pesticides and fuel prices rise in such a way that it ultimately squeezes farmers' incomes. This will lead to decreasing production. 'This could place a double burden upon global agrifood systems: leaving us with poor farmers and lack of food availability at the same time', the FAO Director-General warned. Therefore, in this study, we would like to examine the long-term responsiveness of rice production, in particular to the increase in the relative domestic and international price of rice.

In the literature, the most studied food crops with respect to food security are wheat and maize. However, a considerable portion of the increased demand for these two food crops is coming from animal feed and also biofuels (Baffes & Haniotis, 2016). Therefore, the relationship between farm price and food production from the perspective of food security becomes complicated with these two food crops. Hence, instead of wheat and maize, we have chosen rice as a representative food crop for our analysis. Worldwide, the highest consumers of rice are in G-20 countries, like China, India, Indonesia and Brazil. According to FAO, over 50% of the world population depends on rice for about 80% of its food requirements. A more important note from the point of view of food security is about 95% of the global output of rice is produced and consumed in developing countries, constituting 90% of the world population. Analyzing rice production is also important because of the projected increase in its demand due to increasing population in the Global South (G-20, 2011). FAO (2002) emphasized the renewed efforts in this sector in particular to mitigate the slowdown in the growth of rice production since 1990.

Analyzing the G-20 countries collectively is also interesting due to the heterogeneous nature of this group in terms of its agricultural sector in particular. A significant number of countries in this group are a net importer in terms of agricultural inputs, be it fertilizer, fuel or technology. Hence, a rise in global prices in this account may not encourage production. Also, there exists a significant divergence among the countries in the G-20 group in terms of the volume of rice production and its productivity. Often the large producers from the developing bloc are lagging in terms of productivity relative to their advanced counterparts. Therefore, understanding the increase in productivity and its determining factors including price incentives is also important when discussing food security of the near future.

The rest of study is organized as follows. Section II gives a glimpse of the overall state of agriculture, the problem of food security and rice production in particular in the G-20 countries. Section III is devoted to the literature review, and section IV presents the econometric analysis. The final section concludes.

Agriculture, Food Security and Rice Production in the G-20 Countries

Together, the G-20 members represent two-thirds of the world's population, 60% of all agricultural land and about 80% of world trade in agricultural products. However, among the G-20 countries, there is considerable difference among the members in terms of the importance of agriculture in their economy. According to the World Bank data, for 2022, the percentage share of agricultural value added in the United States accounts for approximately 0.9% of total value added, while that of India is 15.4%. Chinese agriculture contributes to about 7.1% of its gross domestic product (GDP). However, in terms of global agricultural production, the United States ranked next to India; the United States ranked third in 2020 in agricultural output, while India ranked second-highest in global agricultural output. China is the largest grain producer in the world.

Yet, these impressive figures of production do not imply that every citizen of these countries is food secure. The average value of the food security index for the year 2021, measured by the Economists Intelligent Unit, for the developing bloc of the G-20 countries, is 63.4 while that for the developed countries is 78.1. Even though developing countries on average export more food as a percentage of total merchandise export relative to the advanced G-20 countries, among the 113 countries, the average rank of G-20 developing countries is 44 in terms of food availability index (a sub-index of food security) while that of developed countries is 19.[2] The sub-indices also reveal that the developing bloc suffers more from the affordability issue than the availability. It is here noteworthy that within this G-20 group, the developing countries engage around 15% of their labor force in the agricultural sector, while it is only 5.3% for the developed countries. This signifies the need for remunerative prices in agriculture to increase the affordability of this large section of the population. A price incentive can also raise investments in agriculture which by increasing productivity ultimately can address the availability issue. The importance of addressing each and every aspect of food security can be gauged from the fact that over the years, we have seen the problem of food insecurity intensify even in the advanced world. From 2012 to 2021, within the developed bloc of this group, except the United Kingdom, all other major countries registered deterioration in terms of the overall food security index value. India, China and Indonesia have made tremendous progress in this respect, while Brazil and South Africa's decline is noteworthy.

Now, coming to the rice production in particular, the major rice producers in the world are China, India, Indonesia, Bangladesh, Brazil, Thailand, etc. among which China, India, Indonesia and Brazil are part of G-20 countries. The United States also comes in the count of large rice producers and exporters in the world. However, in terms of rice productivity, the top three countries are Australia, Mexico and the United States in that order. Russia also has the same level of productivity as the United States. China and India's average productivity over the last 10 years is, respectively, 4.8 and 2.5 tons per hectare. However, among the 11 countries of the G-20 group, for which we find rice price and output data consistently since 1991, we find a steep rise in both rice production and productivity for almost all the developing countries, e.g. India, Mexico, Turkey, Argentina, etc. However, only in the case of India, we do find a clear domestic price co-movement measured by the rice price index divided by the GDP deflator. For all the other countries, this relative price index for rice either has fallen or follows a slightly upward trend with fluctuations. We find most of the countries' rice prices relative to the global price have increased over time with a downward trend during the first decade of 2000. Argentina and Indonesia witnessed this rise only from 2010. India and Russia faced a secular rise in this regard for the past 30 years. A secular rise of domestic price relative to international price may encourage import or a fall in export opportunity. Both are detrimental to domestic production. From the advanced countries in the G-20 group, we find only Japan to face a downward movement of domestic rice prices relative to the international price. The United States' price, however, fluctuated throughout the study period.

---

[2]According to the World Bank data, for the year 2021, the developing countries' share of food exports as a percentage of total merchandise exports is approximately 16% to 9.3% of the developed countries of the G-20 group. However, in terms of food import as a percentage of total merchandise import, both developed and developing countries' average share is around 8% in this group.

Survey of Literature

Literature solely on the effect of agricultural terms of trade on the output is relatively scarce. However, a large literature can be found on the agricultural supply response to the other factors used directly or even distantly related to agriculture. Schultz (1956) wrote that the quality of inputs is the prime factor behind agricultural growth. However, the nature of agriculture is such that it is inelastic in the short term. Only in the long term, its supply becomes responsive to any changes. After Schultz, the researchers started to give more importance to different constraints that developing economies' farmers face over their price response. However, following Marc Nerlove's seminal work in 1958 in this field, up to the 1970s, there were several studies estimating the short-term and long-term agricultural output responses to the prices. However, Askari and Cummings (1977) were skeptical of the significant positive responses that the different country-level studies found because of their wide differences. In this article, we can find a good survey of applied econometrics literature on the aggregate supply response of agricultural products. Binswanger et al. (1985, 1987) econometrically measured the aggregate supply response in both cross-country and time series frameworks. They found both weak short-term positive and negative supply responses to the prices in time series and cross-country evidence, respectively. They tried to distinguish the effects of price-induced investments in infrastructure and research in agriculture. The study has found significantly high positive effects of price on the output in the long term even though they acknowledged that it is difficult to isolate the price effect only. Binswanger et al. (1989) showed how complex the agricultural output determination process is. Within this process, not only government interact with the farming community, but financial intermediaries also play a crucial role. The district-level study of India showed the importance of government investments and banks in agriculture over and above the prices. Emphasizing the transportation infrastructure, Antle (1983), Judd et al. (1986), Binswanger and Khandker (1995), and Diakosavvas (1990) also conclude in the same direction. Newbery (1992) in arguing for agricultural taxation because of its high responsiveness to public investment showed that the agricultural output is price inelastic. Schiff and Montenegro (1997), however, refused to believe in the average findings of inelastic supply response for the developing economies because of the general consideration of substitutability between price incentives and government investment in agriculture. They argued and empirically showed that these two policies are rather complementary to each other. In this context, FAO (2009) is of the view that even when price incentives exist, developing countries' farmers are not able to reap the benefits and respond accordingly to increase their production due to other supply bottlenecks like lack of credit, infrastructural facilities, etc. Also, in most cases, there exists a significant divergence between international and domestic prices. G-20 (2011), however, stated that in the 1990s, the low commodity prices combined with growing costs of production and energy led to limited supply expansion in the field of agriculture. It also resulted in low investment in exploration and production which only reversed with the commodity price boom in the early 2000. Headey (2018) using ordinary least square (OLS) and generalized method of moments (GMM) methodology, contrary to the general belief, found food inflation to be beneficial for poverty reduction in developing countries because of high supply and wage response to increases in food prices. Headey and Martin (2016) based on different methodologies conclude the same.

For the Chinese economy, Yu et al. (2012) concluded a positive supply response to fluctuations in prices. However, increasing the productivity due to land scarcity, effectiveness of non-price government interventions are also highlighted. In the Indian context, Singh (1995) advocated for improved terms of trade to achieve higher agricultural growth. Palanivel (1995) found a significant positive effect of terms of trade improvement on output for the Indian agricultural economy. At the same time, he underlined the importance of other non-price government support necessary for the structural change in Indian agriculture. Acharya (1997) emphasized the importance of remunerative prices for overall agricultural growth required in the modern era in discussing the Government of India's agricultural policy. Deb (2005) using a cointegrating framework found the Indian agricultural output to be responsive to terms-of-trade changes in the long term if combined with the irrigation facility. Dev and Rao (2010) in discussing the agricultural price support policy of the Government of India acknowledged the importance of support prices in enhancing the supply, investments and technology adoption in agriculture. However, higher food prices are not supportive of poverty reduction or the problem of food security. They were so advocated for non-price interventions of the government in the agricultural sector. Laha and Sinha (2020) analyzed the supply response of food production to the food price fluctuations for the Indian economy in a simultaneous equation framework and found bidirectional causality between the two.

On the whole, from the literature, we find the price to be a weak incentive for agricultural growth on its own. However, its effect in incentivizing more investments in agriculture and thus achieving higher productivity cannot be denied. Again, most of the studies we find are country-level studies for specific crops. However, taking together the major economies of the world who happened to be the major food producers also and then examining their output responses are deficient in number. To the best of our knowledge, the relationship between food price and food production for the G-20 countries is still unexplored. Therefore, as our first attempt, we would like to examine the long-term behavior of all the major rice producers of G-20 countries together in a panel cointegration (and autoregressive distributive lag [ARDL] when applicable) framework.

Econometric Analysis

To identify the trends and patterns that span across multiple time series variables, we have used the panel cointegration analysis which improves the statistical robustness of the results by considering multiple time series data together. This reduces the risk of drawing false conclusions based on individual series. It allows for capturing cross-sectional dependence in the data, which is essential when dealing with panel data where observations are collected from multiple entities over time.

Data and Methodology

We have used yearly data from 1993 to 2021 for the following 11 countries of the G-20 countries: Argentina, Australia, Brazil, India, Indonesia, Japan, Korea, Mexico, Russia, Türkiye and the United States. Based on the availability of data, we have chosen the study period and the set of countries for our study. We have to exclude China because of the non-availability of domestic rice price data for it. The World Bank, the FAO and the Organization for Economic Cooperation and Development (OECD) are the main sources of data for our study.

We have used rice production tonnes per hectare (RPTPH) as a measure of productivity, rice production in 1,000 tonnes (RPTT) as the rice production, producer rice price indices relative to world rice price indices (RPWP) and producer rice price indices relative to GDP deflator (RPGDPDef) as two alternative price indices.[3] As control variables, we have used fertilizer consumption in kilograms per hectare (FCKPH) as a measure of technology inclination and gross fixed capital formation (GFCF) (in agriculture, forestry and fishing) relative to value added (in agriculture, forestry and fishing) as an investment in the agricultural sector and producer support (PS) as a measure of government subsidy.

Panel Unit Root Tests

As we are dealing with time series data of 29 years for only 11 cross-section units, we should treat the variables as time series, and hence, performing the first diagnostic test of the series that is the unit root test is a must. Individual unit root test exercises have a power issue if time series data are present throughout the cross-sections, which might provide erroneous regression findings. To solve this issue, a panel unit root test gives outcomes with greater power. Depending on whether the regression coefficients are homogeneous or heterogeneous, there are two methods for testing panel unit roots. One is Levin et al. (2002) tests where the null hypothesis is a common or homogenous unit root process, others are Im et al. (2003), ADF-Fisher chi-square and PP-Fisher chi-square of Maddala and Wu (1999) and Choi (2001) tests where the null hypothesis is individual or heterogenous unit root process. We perform all the four tests.

Cointegration Test

Cointegration analysis is used to ascertain whether a long-term relationship exists between/among non-stationary variables. The following tests are applicable where all the variables are integrated in the same order. In our case, the variables are

[3]Choosing the appropriate price index to measure the terms of trade has important implications. In the literature, authors have used different indices in this regard. We have used the GDP deflator first due to its availability and comparability to different countries and also to understand the producer price movement relative to such a price index, which includes all prices that farmers consume and also pay for their inputs.

mostly $I(1)$. Usually, two tests are conducted for this purpose. Pedroni (2004) and Kao (1999) provided system-based cointegration for the entire panel through their residual-based cointegration tests, which are both one-way and derived from the Engle–Granger two-step test and Fisher Johansen cointegration test. Maddala and Wu (1999) addressed Fisher's (1932) recommendation to combine the trace test and max-eigen statistics to test for cointegration in a whole panel by combining separate cross sections for cointegration using the Johansen (1988) cointegration test. Johansen (1988) suggested two distinct statistics to ascertain the existence of cointegration vectors in non-stationary time series data: maximal eigenvalue statistics and likelihood ratio trace statistics.

Pedroni provides a number of cointegration tests that take into account heterogeneous trends and intercept coefficients across cross-sections. We have considered the following equation with intercept

$$y_{it} = \alpha + \beta X_{it} + \mu_{it} \tag{10.1}$$

where y_{it} and X_{it} is the cross-sectional dependent and independent variable for unit i at time t. α is the intercept term, β is a coefficient vector and ϵ_{it} is the error term. Under the null hypothesis of no cointegration in the heterogeneous panel, the estimated residual is $I(1)$. Using the above equation, we get the residual and then conduct auxiliary regression for each cross-section to see whether the residual is $I(0)$.

$$e_{it} = \gamma_i e_{it-1} + \sum_{j=1}^{p} \varphi_{ij} \Delta \epsilon_{it-j} + \epsilon_{it} \tag{10.2}$$

Pedroni developed seven test statistics based on the estimated error term e_{it} for testing the null hypothesis of no cointegration ($\gamma_i = 1$). There are two alternative hypothesis: one is homogeneous (within group) hypothesis ($\gamma_i = \gamma$) < 1 for all i, and another one is heterogenous (between group) hypothesis ($\gamma_i < 1$) for all i.

Vector Error Correction Model (VECM)

After confirming the cointegration, the next step is to use the dynamic panel causation test to determine the direction of causation. A panel vector autoregressive (VAR) equation is calculated when cointegration is absent. Only short-term coefficients are computed and short-term causation is inferred in a panel VAR without considering the error correction term (ECT). The panel VAR/VECM Granger causality approach is expected to yield four categories of results: (i) joint causality, in which the ECT coefficient is negative and significant; (ii) short-term causality, in which the coefficients of the short-term explanatory variables are statistically significant; (iii) long-term causality, in which the coefficients of the long-term explanatory variables are statistically significant; and (iv) strong causality, in which the coefficient of both long-term and short-term explanatory variables is statistically significant, and ECT is negative and significant.

The panel VECM contains two steps. First, the long-term relationship is estimated by the following equation:

$$y_{it} = \alpha_i + \sum_{i=1}^{N} \beta_i X_{it} + \epsilon_{it} \tag{10.3}$$

for obtaining the estimated residual ϵ_{it} which is an ECT in the VECM model, where $i = 1, 2, \ldots, N$ and $t = 1, 2, \ldots, T$. The second step is the estimated panel VAR model with dynamic ECT. A panel VEC model can be written as:

$$\begin{aligned} \Delta y_{it} &= \alpha_i + \lambda_i \mathrm{ECM}_{it-1} + \sum_{k=1}^{h} \phi_{ik} \Delta y_{it-k} + \sum_{k=1}^{h} \gamma_{1ik} \Delta X_{1it-k} \\ &+ \sum_{k=1}^{h} \gamma_{2ik} \Delta X_{2it-k} + \cdots\cdots + \sum_{k=1}^{h} \gamma_{nik} \Delta X_{nit-k} + \epsilon_{it} \end{aligned} \tag{10.4}$$

where $i = 1, 2, \ldots, N$ and $t = 1, 2, \ldots, T$. Δ denotes the first difference in the variables, h is the optimal lag length. ECM_{it-1} is the serially uncorrelated ECT, n is the number of exogenous variables in the model, ϕ and η are the coefficients of the estimated parameters. The speed of adjustment, or coefficient of the ECTs, λ_i, indicates how quickly changes in each variable lead to the elimination of deviations from the long-term equilibrium. We tested H_0: $\lambda_i = 0$ for the long-term relationship for all i. For example, if H_0: $\lambda_i = 0$ that implies there is no response from y to a change from the long-term equilibrium in the preceding period.

ARDL Model

The ARDL model was developed by Pesaran et al. (1997, 2001) to estimate the long-term relationship among the variables which are integrated of different orders, i.e. a combination of both $I(0)$ or $I(1)$. The ARDL model is reliably used on short sample periods, and it can be used to distinguish between short-term and long-term coefficients.

The ARDL equation for the panel is shown as follows:

$$\begin{aligned} \Delta \mathrm{RPTT}_{it} &= \gamma_i \left(\mathrm{RPTT}_{i,t-1} - \beta_i^{'} X_{i,t} \right) + \sum_{j=1}^{p} \tau_{ij} \Delta \mathrm{RPTT}_{i,t-1} \\ &+ \sum_{j=0}^{q} \delta_{ij} \Delta X_{i,t-1} + \alpha_i + \epsilon_{it} \end{aligned} \tag{10.5}$$

$$\begin{aligned} \Delta \mathrm{RPTPH}_{it} &= \gamma_i \left(\mathrm{RPTPH}_{i,t-1} - \beta_i^{'} X_{i,t} \right) + \sum_{j=1}^{p} \tau_{ij} \Delta \mathrm{RPTPH}_{i,t-1} \\ &+ \sum_{j=0}^{q} \delta_{ij} \Delta X_{i,t-1} + \alpha_i + \epsilon_{it} \end{aligned} \tag{10.6}$$

where γ_i is the group-specific speed of adjustment coefficient to the long-term equilibrium is represented. In case $\gamma_i = 0$, then there is no indication of existence of a long-term relationship. γ_i is expected to be negative if variables return to a long-term equilibrium. $\beta_i^{'}$ is our vector of interest, indicating how the explanatory factors affect the RPTT over the long term. $(\mathrm{RPTT}_{i,t-1} - \beta_i^{'} X_{i,t})$ and $(\mathrm{RPTPH}_{i,t-1} - \beta_i^{'} X_{i,t})$ is the ECT. τ_{ij} and δ_{ij} are the short-term dynamic coefficients. Δ indicates first difference operator. p and q are optimal lag order of the dependent and independent variables. α_i denotes the fixed effects. ϵ_{it} is the error term and is assumed to be white noise that varies across time and countries.

Results and Discussion

Table 10.1 describes the raw data. We can observe that there is significant variation in rice production (as expected) and in producer support. The fertilizer use and investments also vary across units over the periods.

We find a good correlation between rice productivity and domestic relative rice price and investments in agriculture (0.50 and 0.58, respectively). Table 10.1 reveals producer support is also increasing with the domestic relative price of rice (value of correlation is 0.52). It also helps in more fertilizer consumption (correlation is 0.63). The important message to note here is that rice productivity and rice production are negatively correlated. The rice production except fertilizer consumption is related to all the other variables negatively though very weakly.

Table 10.2 shows the panel unit root test results where we have found RPTPH, RPWP, FCKPH and GFCF are integrated of order one (*I*(1), non-stationary variables) and RPTT, PS and the RPGDPDef are integrated of order zero (*I*(0), stationary variables). Therefore, to analyze the long-term relation among the variables, we have used both the panel cointegration technique for the *I*(1) variables and the ARDL model for *I*(0) and *I*(1) variables.

The Pedroni, Kao and Fisher-Johansen test methodologies are used here for conducting the panel cointegration test. The Pedroni approach test also has three distinct criteria: 'no deterministic trend', 'deterministic intercept and trend' and 'no deterministic intercept and trend'. Once more, each criterion has two subcategories: between dimension (which includes three group statistics) and within dimension (which covers four-panel statistics). Table 10.2 presents the results of the Pedroni residual cointegration test. Based on our graphical observations, we have determined that there is no deterministic trend under this test. The results confirmed that panel series for RPPTH, RPWP and FCKPH are cointegrated over the years 1993–2021. As the GFCF data are available from 2001 only, we have separately checked and confirmed that all the *I*(1) panel series including GFCF are cointegrated for the period 2001 to 2021.

We also present the Kao cointegration test and Fisher cointegration test result in Table 10.2. The analysis indicates that there is a long-term relationship among the series in the panel data framework since the residual of the panel RPPTH upon RPWP and FCKPH is stationary at level and the probability value of the ADF test statistic is significantly lower than the intended value of 0.10 over the years 1993–2021. Including GFCF for 2001 to 2021 also confirms the result.

The results of the Fisher-Johansen test show for the periods 1993 to 2021 and 2001 to 2021, there is one cointegrating equation each among RPPTH, RPWP, FCKPH and RPPTH, RPWP, FCKPH, GFCF, respectively. By integrating the three results, we can infer that panel series are cointegrated and have an overall equilibrium connection.

In the lower section of Table 10.2, we represent the panel error correction result which shows that the sign of the ECTs is negative and significant. This means any errors in the short term are being corrected. VECM results confirm that there is a long-term equilibrium relationship running from RPWP, FCKPH and GFCF to RPTPH. However, here we should note that RPWP is related to

Table 10.1. Descriptive Statistics and Pair-Wise Correlation of Among Variables for 11 G-20 Members.

	RPTT	**RPTPH**	**RPGDPDef**	**RPWP**	**FCKPH**	**GFCF**	**PS**
Mean	15,100.45	4.77	0.82	0.87	150.86	18.45	0.19
Median	4,268.08	4.72	0.89	0.84	125.33	16.27	0.17
Maximum	125,031.60	10.94	2.05	2.76	478.15	56.39	0.87
Minimum	17.61	1.74	0.03	0.11	11.00	2.09	−0.68
Std. Dev.	29,406.98	1.81	0.43	0.46	105.07	10.34	0.32
Skewness	2.48	1.17	−0.10	1.31	0.82	0.73	0.17
Kurtosis	7.94	4.88	2.22	5.58	2.98	3.69	3.11
Observations	230.00	230.00	230.00	230.00	230.00	230.00	230.00
	RPTT	**RPTPH**	**RPGDPDef**	**RPWP**	**FCKPH**	**GFCF**	**PS**
RPTT	1.00						
RPTPH	−0.52***	1.00					
RPGDPDef	−0.16**	0.50***	1.00				
RPWP	−0.15**	0.25***	0.31***	1.00			
FCKPH	0.13***	−0.19***	0.05	0.37***	1.00		
GFCF	−0.14**	0.58***	0.39***	0.24***	−0.04	1.00	
PS	−0.29***	0.13*	0.52***	0.31***	0.63***	0.18**	1.00

Source: Authors' calculation.

Note: *** implies the 1% and ** implies the 5% level of significance.

Table 10.2. Panel Unit Root Test, Panel Cointegration and Long-Term Relationship and VECM Result for 11 G-20 Members.

Panel Unit Root Test for RPPTH, RPWP, FCKPH and GFCF							
Test Statistic With Intercept							
		1993–2021					2001–2021
Methods	Null Hypothesis	RPTT	RPPTH	RPGDPDef	RPWP	FCKPH	GFCF
		I(0)	*I*(1)	*I*(0)	*I*(1)	*I*(1)	*I*(1)
Levin, Lin and Chu t	Unit root (assumes common unit root process)	−1.23 [0.11]	−16.89 [0.00]	−2.34 [0.01]	−10.79 [0.00]	−21.51 [0.00]	−8.68 [0.00]
Im, Pesaran and Shin *W*-stat	Unit root (assumes individual unit root process)	−1.86 [0.03]	−18.91 [0.00]	−3.97 [0.00]	−11.07 [0.00]	−21.39 [0.00]	−9.02 [0.00]
ADF – Fisher Chi-square	Unit root (assumes individual unit root process)	47.87 [0.00]	248.21 [0.00]	59.61 [0.00]	146.97 [0.00]	237.55 [0.00]	124.32 [0.00]
PP – Fisher Chi-square	Unit root (assumes individual unit root process)	49.20 [0.00]	294.13 [0.00]	80.12 [0.00]	144.92 [0.00]	305.18 [0.00]	334.82 [0.00]
Panel Cointegration Test							
Pedroni Residual Cointegration test for RPPTH, RPWP and FCKPH series, 1993–2021							
No deterministic trend							
Null Hypothesis: No Cointegration			Statistic	Prob.	Weighted Statistic		Prob.
Alternative hypothesis Common AR coefficients (within dimension)	Panel *v*-Statistic		0.30	0.38	0.07		0.47
	Panel rho-Statistic		−2.60	0.00	−2.49		0.01
	Panel PP-Statistic		−4.09	0.00	−4.15		0.00

(*Continued*)

Table 10.2. *(Continued)*

Panel Unit Root Test for RPPTH, RPWP, FCKPH and GFCF						
Test Statistic With Intercept						
	Panel ADF-Statistic		−4.03	0.00	−4.19	0.00
Alternative hypothesis:	Group rho-Statistic		−2.53	0.01	–	–
Individual AR coefficients	Group PP-Statistic		−5.69	0.00	–	–
(between dimension)	Group ADF-Statistic		−6.01	0.00	–	–
Kao Residual Cointegration Test for RPPTH, RPWP, FCKPH, 1993–2021						
No deterministic trend						
Null Hypothesis: No cointegration			*t*-Statistic		Prob.	
ADF			1.55		0.06	
Fisher Johansen Cointegration Test for RPPTH, RPWP and FCKPH 1993–2021						
No deterministic trend (restricted constant)						
Null hypothesis: No Cointegrated Equation	Fisher Stat.*(from trace test)	Prob.		Fisher Stat.* (from max-eigen test)		Prob.
None	68.15	0.00		58.25		0.00
At most 1	30.66	0.10		26.02		0.25
At most 2	21.92	0.46		21.92		0.46
Long-Term Relationship and VECM Result, 1993–2021						
Dependent Variables	Independent Variables		EC Terms		Probability	Error Corrected
RPTPH/D(RPTPH)	RPWP(-1)	FCKPH (-1)				
	8.00	0.02	$C(1) = -0.014$		0.00	Yes
	2.27	−7.31				

Source: The author's calculation.

Note: The numbers in [] parentheses are showing the probability value, presented in the upper panel.

the PRTPH inversely in the long term, and the long-term coefficient for FCKPH is not significant at all (see lower panel of Table 10.2). From this result, we can infer that domestic rice prices relative to the international price do not co-move with rice productivity over time. In free market economies, a rise in RPWP creates a disincentive for the domestic producers by encouraging imports and discouraging exports. This finding confirms our casual observations of the rice productivity and the price data discussed in section II where we have seen that rice productivity mainly increased over time for the developing countries of the G-20 group only, and these countries faced on an average a rise in domestic rice prices relative to the international price except the first decade of this millennium. Similarly, fertilizer consumption which is generally higher in the advanced world which in turn does not experience much rise in their productivity can explain the insignificance of its long-term coefficient. Including investments in agriculture does not change the above results qualitatively. Here, one thing we should note is that we do not have fertilizer consumption or investment for rice production only. The data we use are overall fertilizer consumption and investment in agriculture for the countries.

Now, to analyze the effect of domestic relative rice price (RPGDPDef) which is mainly found to be an $I(0)$ variable, we are using the ADRL methodology. Also, we find rice production to be an $I(0)$ variable. Therefore, in Table 10.3, we have presented the ARDL results for different dependent variables. In model 1, the dependent variable is RPTT and the independent variables are RPGDPDef and FCKPH. Here, we found a positive significant long-term relation between the price and output and also between fertilizer consumption and output. The ECM term is also statistically significant with a desirable negative sign implying the long-term stable relation. In model 2 where instead of RPGDPDef we have included RPWP, as expected, only the coefficient for the fertilizer is statistically significant probably for the aforementioned reasons. Interestingly, in model 3 also, which estimates the effect of fertilizer consumption and the domestic relative price of rice on rice productivity, we find fertilizer to be significantly negatively related to rice productivity. This long-term relation is also found to be stable at a 5% level of significance. In this case, no statistically significant effect of the domestic relative price of rice on rice productivity is found.

The lower panel of Table 10.3 shows results for the period 2001 to 2021, where we have included the investment in agriculture as an additional control variable. However, this does not have any impact on the relationship between rice production and the domestic relative price of rice. However, in models 3 and 4 where instead of the domestic relative price of rice we have taken rice price relative to international price, we found interesting results. Here, RPWP is becoming positively significant at the expense of GFCF. But this relation vanished when instead of GFCF, we included fertilizer consumption. In models 5 to 8 where we have tried to explain the rice productivity including investments from 2001 to 2021, we find the ECTs to be insignificant. Therefore, it will not be wise to rely on the long-term estimates. One thing to be mentioned here is that we cannot include different measures of rice price, investment and fertilizer consumption in one equation due to the problem of the singular matrix. Also, due to the same

Table 10.3. ARDL Results for Different Dependent Variables.

ARDL Result for the Time Period 1991–2021

	Model-1	Model-2		Model-3
	D(RPTT)	D(RPTT)		D(RPTPH)
		Long Run		
RPGDPDef	1163.31**		RPGDPDef	–0.42
	[2.51]			[–0.95]
RPWP		64.45	FCKPH	–0.003***
		[0.72]		[–2.76]
FCKPH	5.22**	6.8***	*Short Run*	
	[5.02]	[5.37]	Coint Eq01	–0.20**
	Short Run			[–2.56]
Coint Eq01	–0.27***	–0.38***	C	1.44
	[–3.08]	[–3.78]		[2.56]
C	1168.98	2555.52***		
	[1.38]	[3.02]		

ARDL Result for the Time Period 2001–2021

	Model-1	Model-2		Model-3	Model-4		Model-5	Model-6		Model-7	Model-8
	D(RPTT)							D(RPTPH)			
					Long Run						
RPGDPDef	3434.18***	1717.43	RPWP	309.59***	–3545.12***	RPGDPDef	0.85***	–0.86***	RPWP	1.24***	189
	[9.13]	[0.83]		[76.58]	[–14.29]		[4.1]	[–2.48]		[20.33]	[0.04]
GFCF	–7.48		GFCF	–50.62***		GFCF	–0.2		GFCF	0.09***	
	[–0.43]			[–29.2]			[–12.1]			[16.67]	
FCKPH		185.91**	FCKPH		61.12***	FCKPH		–0.01***	FCKPH		–0.58
		[2.07]			[26.27]			[–13.02]			[–0.04]
					Short Run						
Coint Eq01	–0.34**	–0.02	Coint Eq01	–0.6***	–0.29	Coint Eq01	–0.38	–0.38*	Coint Eq01	0.05	0
	[–2.45]	[–0.34]		[–2.68]	[–1.39]		[–1.63]	[–1.85]		[0.16]	[1.18]
C	2182.59	4823.63	C	3948.7*	4992.18	*C*	2.74	2.65**	C	0.38	0.33**
	[1.14]	[1.1]		[1.68]	[0.99]		[1.47]	[2.28]		[0.71]	[2.12]

Source: Authors' calculation.

Note: The Numbers in [] parentheses are the values of the *t*-statistic. ***, **, * indicate 1%, 5% and 10% level of significance, respectively.

problem, we have to analyze another control variable, the producer's support separately. As the PP is found to be an *I*(0) variable, we have first tried to analyze its long-term effect in the ARDL framework. However, no long-term relation can be established between the variables namely rice production/productivity, alternative relative prices of rice and producer support. Now, as rice production, the domestic relative price of rice and producer support are all *I*(0) variables, we have tried to simply estimate panel regression with these three variables. Based on the Hausman test, we find the random effect model to be optimal. However, in that case, also, we only find the domestic relative price of rice to be the only significant determinant of rice production (results are not shown here).

Conclusion

From our analysis of rice production and productivity of the G-20 countries, we find that the domestic relative price of rice to be the only significant determinant of rice production. This finding confirms our hypothesis that improvement in food prices increases food production. Therefore, rising food prices do not univocally cause global food insecurity. Here, one important point to note is that we do not find any significant and consistent effect of domestic rice price relative to international rice price on the rice production. However, a higher domestic price relative to the world is discouraging productivity possibly, for open economies like G-20 countries, as it encourages more imports or reduces the export opportunities.

None of the other variables such as investment in agriculture, overall technological development in agriculture captured through fertilizer use in agriculture or overall government support have any impact on either the production or productivity of rice. This signifies the importance of FAO's concern over the declining rice productivity at the global level (FAO, 2002). Our econometric analysis as well as observation of raw data suggests that even the advanced countries which attract a huge amount of government support and investments in agriculture relative to the developing countries do not show a commensurate increase in productivity or production during the last 30 years. Somehow, its production and productivity have stagnated even with a high technological inclination. This must bear a serious concern over the long term as it is projected that the world is going to face a huge increase in the requirement of rice due to the projected growth in population. Therefore, developing countries are suggested to focus more on their productivity effort to bridge this gap.

References

Abbott, P. C., Hurt, C., & Tyner, W. E. (2008). What's driving food prices? Report No. 37951. *Farm Foundation.* http://dx.doi.org/10.22004/ag.econ.37951

Acharya, S. S. (1997). Agricultural price policy and development: Some facts and emerging issues. *Indian Journal of Agricultural Economics.* https://doi.org/10.22004/ag.econ.297511

Antle, J. M. (1983). Infrastructure and aggregate agricultural productivity: International evidence. *Economic Development and Cultural Change.* http://www.jstor.org/stable/1153216

Askari, H., & Cummings, J. T. (1977). Estimating agricultural supply response with the Nerlove model: A survey. *International Economic Review.* https://doi.org/10.2307/2525749

Baffes, J., & Haniotis, T. (2010). Placing the recent commodity boom into perspective. In A. Aksoy & B. Hoekman (Eds.), *Food prices and rural poverty.* World Bank Group. http://documents.worldbank.org/curated/en/612501468154488885/Food-prices-and-rural poverty

Baffes, J., & Haniotis, T. (2016). What explains agricultural price movements? *Journal of Agricultural Economics.* https://doi.org/10.1111/1477-9552.12172

Binswanger, H. P., & Khandker, S. R. (1995). The impact of formal finance on the rural economy of India. *Journal of Development Studies*. https://doi.org/10.1080/00220389508422413

Binswanger, H. P., Khandker, S. R., & Rosenzweig, M. R. (1989). *How infrastructure and financial institutions affect agricultural output and investment in India*. Policy Research Working Paper Series 163. The World Bank. http://www-wds.worldbank.org/external/default/WDSContentServer/WDSP/IB/1989/03/01/000009265_3960927171152/Rendered/PDF/multi_page.pdf

Binswanger, H., Mundlak, Y., Yang, M., & Bower, A. (1985). *Estimation of aggregate agricultural supply response from time series of cross-country data*. Agricultural Research Unit, discussion paper no. ARU48. World Bank Group. http://documents.worldbank.org/curated/en/137311468740375928/Estimation-of-aggregate-agricultural-supply-response-from-time-series-of-cross-country-data

Binswanger, H., Yang, M., Bowers, A., & Mundlak, Y. (1987). On the determinants of cross-country aggregate agricultural supply. *Journal of Econometrics*. https://doi.org/10.1016/0304-4076(87)90046-7

Brümmer, B., Korn, O., Schlüßler, K., Jaghdani, T. J., & Saucedo, A. (2013). *Volatility in the after crisis period? A literature review of recent empirical research*. Working Paper 1. Department of Agricultural Economics and Rural Development University of Göttingen. http://www.fp7-ulysses.eu/

Choi, I. (2001). Unit root tests for panel data. *Journal of International Money and Finance*, *20*, 249–272. https://doi.org/10.1016/S0261-5606(00)00048-6

Clapp, J., & Cohen, M. J. (2009). The food crisis and global governance. In *The global food crisis: Governance challenges and opportunities*. http://www.scopus.com/inward/record.url?eid=2-s2.0-77953814346&partnerID=MN8TOARS

Deb, S. (2005). Terms of trade and supply response of Indian agriculture: Analysis in cointegration framework. *Indian Economic Review*. http://www.jstor.org/stable/29793832

Dev, S. M., & Rao, N. C. (2010). Agricultural price policy, farm profitability and food security. *Economic and Political Weekly*. http://www.jstor.org/stable/40736698

Dewbre, J., Giner, C., Thompson, W., & Lampe, M. V. (2008). High food commodity prices: Will they stay? Who will pay? *Agricultural Economics*. https://doi.org/10.1111/j.1574-0862.2008.00346.x

Diakosavvas, D. (1990). Government expenditure on agriculture and agricultural performance in developing countries: An empirical evaluation. *Journal of Agricultural Economics*. https://doi.org/10.1111/j.1477-9552.1990.tb00654.x

Elliott, K. A. (2008). *Biofuels and the food price crisis: A survey of the issues*. Center for Global Development Working Paper No. 151. http://dx.doi.org/10.2139/ssrn.1221668

Fisher, R. A. (1932). Inverse probability and the use of likelihood. *Mathematical Proceedings of the Cambridge Philosophical Society*, *28*, 257–261. https://doi.org/10.1017/S0305004100010094

Food and Agriculture Organization. (2002). *FAO rice information*. https://www.fao.org/3/Y4347E/y4347e00.htm

Food and Agriculture Organization. (2009). *High food prices and the food crisis – Experiences and lessons learned*. https://www.fao.org/3/i0753e/i0753e00.htm

Food and Agriculture Organization. (2022). *G20: FAO urges Finance Ministers and Central Bank Governors to support the food import financing facility*. https://www.fao.org/newsroom/detail/g20-fao-urges-finance-ministers-and-central-bank-governors-to-support-the-food-import-financing-facility/en

G-20. (2011). *Ministerial declaration, action plan on food price volatility and agriculture.* Meeting of G20 Agriculture Ministers. http://reliefweb.int/sites/reliefweb.int/files/resources/Full_Report_5.PDF

García-Germán, S., Bardají, I., & Garrido, A. (2015). Evaluating price transmission between global agricultural markets and consumer food price indices in the European Union. *Agricultural Economics.* https://doi.org/10.1111/agec.12209

Gilbert, C. L. (2010). How to understand high food prices. *Journal of Agricultural Economics, 61*(2), 398–425. https://doi.org/10.1111/j.1477-9552.2010.00248.x

Gouel, C. (2012). Agricultural price instability: A survey of competing explanations and remedies. *Journal of Economic Surveys.* https://doi.org/10.1111/j.1467-6419.2010.00634.x

Haile, M. G., Kalkuhl, M., & von, B. J. (2014). Agricultural supply response to international food prices and price volatility: A cross-country panel analysis. In *2014 International Congress.* https://doi.org/10.22004/ag.econ.182725

Headey, D. D. (2018). Food prices and poverty. *The World Bank Economic Review.* https://www.jstor.org/stable/48557126

Headey, D., & Fan, S. (2008). Anatomy of a crisis: The causes and consequences of surging food prices. *Agricultural Economics.* https://doi.org/10.1111/j.1574-0862.2008.00345.x

Headey, D. D., & Martin, W. J. (2016). The impact of food prices on poverty and food security. *Annual Review of Resource Economics.* https://www.jstor.org/stable/26773369

Heltberg, R., Hossain, N., & Reva, A. (2012). *Living through crises: How the food, fuel and financial shocks affect the poor* (Report No. 67865). World Bank Publications. http://documents.worldbank.org/curated/en/437601468330872450/Living-through-crises-how-the-food-fuel-and-financial-shocks-affect-the-poor

Im, K. S., Pesaran, H., Shin, Y., & Smith, R. J. (2003). Testing for unit roots in heterogeneous panels. *Journal of Econometrics, 115,* 53–74.

Johansen, S. (1988). Statistical analysis of cointegrating vectors. *Journal of Economic Dynamics and Control, 12,* 231–254. https://doi.org/10.1016/0165-1889(88)90041-3

Judd, M. A., Boyce, J. K., & Evenson, R. E. (1986). Investing in agricultural supply: The determinants of agricultural research and extension investment. *Economic Development and Cultural Change.* http://www.jstor.org/stable/1154145

Kao, C. (1999). Spurious regression and residual-based tests for cointegration in panel data. *Journal of Econometrics, 90,* 1–44. https://doi.org/10.1016/S0304-4076(98)00023-2

Krugman, P. (2008). 'Grains gone wild', Op-Ed. *New York Times.* https://www.nytimes.com/2008/04/07/opinion/07krugman.html

Laha, A., & Sinha, S. (2020). Implications of food price shocks on availability of food: Evidences from the Indian economy. *Millennial Asia.* https://doi.org/10.1177/0976399620937399

Levin, A., Lin, C. F., & Chu, C. S. J. (2002). Unit root tests in panel data: Asymptotic and finite-sample properties. *Journal of Econometrics, 108,* 1–24. https://doi.org/10.1016/S0304-4076(01)00098-7

Maddala, G. S., & Wu, S. (1999). A comparative study of unit root tests with panel data and a new simple test. *Oxford Bulletin of Economics and Statistics, 61,* 631–652. https://doi.org/10.1111/1468-0084.61.s1.13

Newbery, D. M. (1992). Agricultural pricing and public investment. *Journal of Public Economics*. https://doi.org/10.1016/0047-2727(92)90050-P

Organization for Economic Cooperation and Development and Food. (2008). *Rising agricultural prices: Causes, consequences and responses*. https://www.europeansources.info/record/policy-brief-rising-agricultural-prices-causes-consequences-and-responses/

Organization for Economic Cooperation and Development and Food and Agriculture Organization. (2008). *OECD-FAO Agricultural Outlook 2008–2017*. https://www.globalbioenergy.org/uploads/media/0807_OECD-FAO_-_Agricultural_Outlook_2008-2017.pdf

Palanivel, T. (1995). Aggregate supply response in Indian agriculture: Some empirical evidence and policy implications. *Indian Economic Review*. http://www.jstor.org/stable/29793722

Pedroni, P. (2004). Panel cointegration: Asymptotic and finite sample properties of pooled time series tests with an application to the PPP hypothesis. *Econometric Theory*, *20*, 597–625.

Pesaran, M. H., Shin, Y., & Smith, R. P. (1997). *Pooled estimation of long-run relationships in dynamic heterogeneous panels*. University of Cambridge, Department of Applied Economics, Cambridge.

Pesaran, M. H., Shin, Y., & Smith, R. (2001). Bounds testing approaches to the analysis of level relationships. *Journal of Applied Econometrics*, *16*, 289–326. https://doi.org/10.1002/jae.616

Schiff, M., & Montenegro, C. E. (1997). Aggregate agricultural supply response in developing countries: A survey of selected issues. *Economic Development and Cultural Change*. http://www.jstor.org/stable/1154541

Schultz, T. W. (1956). Reflections on agricultural production, output and supply. *Journal of Farm Economics*. https://doi.org/10.2307/1234459

Singh, M. (1995). Inaugural address at the 54th annual conference of the Indian society of agricultural economics. *Indian Journal of Agricultural Economics*, *50*(1), 1–6.

Tadesse, G., Algieri, B., Kalkuhl, M., & Von Braun, J. (2014). Drivers and triggers of international food price spikes and volatility. *Food Policy*. https://doi.org/10.1016/j.foodpol.2013.08.014

Wolf, M. (2008). *Food crisis is a chance to reform global agriculture*. Financial Times. https://www.ft.com/content/2e5b2f36-1608-11dd-880a-0000779fd2ac

World Bank. (2008). *Rising food prices: Policy options and World Bank response* (Report No. 43225). World Bank Group. http://documents.worldbank.org/curated/en/990671468313502111/Rising-food-prices-policy-options-and-World-Bank-response

Wright, B. D. (2012). *International grain reserves and other instruments to address volatility in grain markets*. The World Bank Research Observer. https://doi.org/10.1093/wbro/lkr016

Yu, B., Liu, F., & You, L. (2012). Dynamic agricultural supply response under economic transformation: A case study of Henan, China. *American Journal of Agricultural Economics*. http://www.jstor.org/stable/41331261

Chapter 11

Has Indian Agricultural Growth Manage Food Security in India Successfully in the Last Decade? An Empirical Investigation

Tanushree Gupta[a], *Jennifer Fernandes*[b] *and Susobhan Maiti*[b]

[a]JSPM University, India
[b]JAIN (Deemed-to-be University), India

Abstract

Food security emerges when all people always have physical and economic access to sufficient, safe, and nutritious food to meet their dietary needs and food preferences for an active and healthy life. The availability of food, access to food, and absorption of food are three key aspects of food security. Considering the above, the initial goal in this chapter is to measure annual growth rate and compound annual growth rate (CAGR) of production of food grains, nutri coarse cereals, total pulses, and nine oilseeds for the period 2011–2021 in India. To estimate per capita availability of Food Grains, Nutri Coarse Cereals, Total Pulses and Nine Oilseeds (in kg) represents the food security in India. The highest production is recorded in between 2011 and 2021 for food grains (315.72), followed by the production of nine oil seeds (36.7), total pulses (27.69), and nutri coarse cereals (26.42). In 2016, the annual growth rate for food grains and nine oilseeds was 9.37 and 23.88%, respectively. It is very evident that the availability per capita has grown from 2011 to 2021. The per capita availability of food grains increased from 206.17 kg in 2011 to 224.30 kg in 2021. The per capita availability of total pulses increased from 13.59 kg in 2011 to 19.67 kg in 2021. The per capita availability of nine oil seeds increased from 23.70 kg in 2011 to 26.78 kg in 2021 which indicates India has managed food security successfully in the last decade.

Keywords: Indian agriculture; agricultural growth; food security; per capita food availability; food grains

Growth, Poverty and Developmental Aspects of Agriculture, 163–176

doi:10.1108/978-1-83608-076-320241011

Introduction

An active life could be supported by food security, according to the World Food Summit (WFS) and the Food and Agricultural Organization (FAO). This is achieved when all people at all times have physical and economic access to sufficient, safe, and nutritious food that meets their dietary and food preferences (Pillay & Kumar, 2018). Additionally, our bodies obtain fuel, in the form of energy, from food and nourishment. Inadequate nutrition has long-lasting effects on both individuals and society as a whole (NABARD, 2022).The concept of food security was first proposed in the 1970s during the World Food Conference, where it was described as the assurance of both the availability and the price stability of food. This meant that there should be sufficient supplies of basic foodstuffs worldwide to support a constant increase in food consumption, even in the face of price and production fluctuations. As far as food security is concerned, three vital aspects are: the availability of food, access to food, and absorption of food. On the contrary, the World Bank introduced the concepts of "chronic" and "transitory" food insecurity in their 1986 Poverty and Hunger Report. The former is linked to low incomes and structural poverty, while the latter is linked to natural disasters, economic collapse, or conflicts. Agricultural development is essential for enhancing food security, as it will boost food production and availability right away, although a guarantee of food security requires more than just the availability of food (Swaminathan et al., 2014).

Since gaining independence in 1947, India has implemented agricultural development strategies with the goal of swiftly diminishing hunger, food insecurity, malnutrition, and poverty. Despite the latest worldwide food crisis – which has sparked political and social turmoil in several developing world nations – India's agricultural development strategy and approach to food security has once again demonstrated its resilience. As early as the mid-1960s, India's agricultural development plan began to center on ensuring food security on a national and household level. The new strategy focused on establishing a strong basis for food security in India through three essential components: (a) providing farmers with an enhanced technology package, (b) delivering modern agricultural inputs, technical expertise, and institutional credit to farmers, and (c) ensuring a profitable marketing and pricing environment for farmers. It is necessary to evaluate a country's food security on three different levels. First, there must be a sustained supply of food on a national level, which is dependent on either an appropriate capacity to import food or an adequate level and increase of food production. Second, every home should be able to physically get food. A well-organized system of advertising, transportation, and storage must be in place to provide food to people's doorsteps or within an acceptable distance from their homes. Every family's ability to buy food is influenced by their budget and the pricing of the food that is readily available to them. Finally, the third factor is how people use the food that is available to them; this is dependent on how food is distributed within families and on everyone staying healthy enough to eat and absorb the necessary amount of food (Acharya, 2009; Pawlak & Kołodziejczak, 2020).

The Indian economy has been expanding at a phenomenal rate recently, and it is still among the world's fastest-growing countries. Despite numerous advancements, poverty and food insecurity continue to be issues in India. Food is a fundamental necessity for human survival, development, and maturation.

According to the Global Hunger Index for 2021, out of 116 nations, India came in at number 101. The FAO reports that between 2020 and 2021, food prices rose by 30%. Even though the Indian government has long worked to ensure that families have access to nutritious food through programs like the Public Distribution System (PDS) and the National Food Security Act (NFSA) 2013, there are still pressing issues surrounding food security in India due to factors like a growing population, changing weather patterns, and disruptions in global supply chains caused by the conflict between Russia and Ukraine(Food Security in India, 2022).

India addressed food insecurity and hunger with long-term and short-term initiatives. It developed a long-term development strategy to maintain enough food production. Apart from them, selective market intervention and targeted distribution of subsidized food reduced hunger and food insecurity in the near run. Nonfood issues like clean water and sanitation also affect nutritional status; therefore, addressing them can help food security become good nutrition. The biggest shift is the rise in cereal production. However, price support for farmers has been a significant tool for agricultural development and food policy since the mid-1960s. Cereal minimum support prices are key price policy tools. Providing food subsidies and key farm supplies to farmers at fair costs or user charges helps. Cereals, pulses, oilseeds, raw cotton, raw jute, and sugarcane are among 25 agriculture products with minimum support prices. Edible oil and pulse imports grow significantly after this (Acharya, 2009).

With respect to the above background, this chapter aims to investigate to measure annual growth rate and compound annual growth rate (CAGR) of production of food grains, nutri coarse cereals, total pulses, and nine oilseeds for the period 2011–2021 in India. Secondly, to estimate per capita availability of food grains, nutri coarse cereals, total pulses, and nine oilseeds for the period 2011–2021 which may represent the food security in India in the last decade.

Review of Literature

Since 1947, India has gone through four phases in agricultural evolution; Phase I (1947–1964) focuses on building the infrastructure for scientific agriculture, which included starting agricultural universities, organizing national extension and community development programs, building large irrigation and power projects, and establishing factories for fertilizer and pesticides. Phase II (1965–1985) emphasis on maximizing the benefits of the infrastructure created during Phase I, particularly in irrigation and technology transfer. Phase III (1985–2000): Organizations' innovations were introduced, but governmental spending on irrigation and other infrastructure necessary for agricultural development gradually decreased, and the cooperative credit system gradually broke down. Grain reserves contributed to a complacent attitude toward agriculture. Phase IV (2001 onwards): Policymakers' weariness led to a stagnation in output, technology, and extension. Little progress was made in altering the policies pertaining to inputs, irrigation, food pricing, subsidies, and distribution because of a weak coalition government and strong lobby interests (Marothia et al., 2022).

However, recent initiatives, such as the Mahatma Gandhi National Rural Employment Guarantee Act (MGNREGA), aim to halt this decline. Policies to improve agricultural storage and rectify the gap between production and post-harvest technology are also being explored (Swaminathan et al., 2014). India's food production has increased significantly because of agricultural growth, although chronic undernutrition still exists. Focusing on livestock and crops that have significant nutritional effects on farmers and consumers is perhaps one of the solutions, and addressing the socioeconomic issues that impact the relationship between agriculture and nutrition, such as the distribution of assets, notably land, women's roles, rural infrastructure, and rural health and sanitation, may be another component (Gillespie & Harris, 2012).

India accounts for 2.4% of the world's landmass. States in India have an average landholding size of 1.08 hectares. Of these, half have marginal farmers (land holdings of less than 1 ha), and the other half have small farmers (land holdings of 1–2 ha). Availability of credit, the supply of inputs, appropriate transportation, and market facilities are some of the major constraints faced by the farmers (Agricultural Census, 2015–2016). Due to large yearly variations in monsoon rainfall, India's agriculture is far riskier than China's, with roughly 60% of its cultivated area still being rainfed. Significant decreases in volatility have been observed in both countries as a result of increased irrigation, the adoption of more recent, robust technology, and diversification. India's agricultural growth is still far more erratic than China's, though. Periodic weather-related interruptions continue to be a major problem for Indian agriculture and are probably a major factor in the decisions farmers make about their investments. In order to stabilize their overall sector growth rates, China and India have both made significant irrigation investments; but China's growth has been comparatively more stable due to its faster diversification into noncereal crops and, in particular, noncrop agriculture (Gautam & Yu, 2015).

Agriculture promotes economic change and development in India through its causal links with factor and product markets. Increasing agricultural productivity promotes growth and offers a means of reducing poverty. States all around the country saw dismal agricultural growth performance. From 2011 to 2017, a significant number of states had either negative or less than 1% increase in gross value added (GVA) at 2011–2012 prices. A bottom-up strategy, similar to Kerala's people campaign model, is required to effect change. Collectives like self-help organizations should be encouraged in order to provide farmers more negotiating power in the marketplace and to increase the effectiveness of delivery systems by giving them more authority. Enhancing land production through total factor productivity which is fueled by technology is crucial. Income security for farmers could be achieved by diversifying their household's revenue streams through nonfarm pursuits (Krishi & Raipur, 2017). Due to an inefficient use of resources, production rates in the agricultural sector in India have been dropping. India should implement contract farming to help lower risks and encourage farmers to keep up their output. Increased public spending on irrigation, purchasing power parity (PPP) in agriculture, and research and development (R&D) investments in agriculture are essential. Consolidating holdings is necessary since India's land is extremely fragmented, with 80% of farmers falling into the small and marginal farmer group.

The best way to encourage farmers to increase their agricultural production investments would be to give them access to finance and competitive pricing for their produce (Saatwik Panigrahi, 2020). Although the green revolution helped alleviate part of the food crisis that arose in the early 1960s, it also brought with it the regional differences in cropping patterns, productivity, and resource consumption.

Certain cash crops experience scarcity as a result of promotional pricing policies. The degree of vulnerability in farm earnings and farmers' livelihoods overall is rising due to changes in environmental variables, economic factors, and technological factors (Singh, 2020). Being the primary source of income for over 80% of Jharkhand's population, agriculture plays a crucial role in both the state's overall economic growth and its efforts to end poverty. Because there is little irrigation and most farming is rainfed, the state's agriculture is mostly dependent on the monsoons. Since 2000, there has been a major shift in Jharkhand's cropping patterns toward oilseeds, maize, and pulses. The value of agricultural output increased at a rate of about 8% annually during this period, although with significant annual volatility. The development of appropriate technologies for rained farming, short-duration oilseed crops, like safflower, sesame, and linseed, as well as pulses, like green gram, black gram, pigeon pea, and Bengal gram, should be encouraged because they are well-suited to the area's light, shallow soil cover. High-value commodities, which can be a sustainable source of growth and act as a buffer for agricultural expansion, should be prioritized while diversifying agricultural endeavors. Therefore, government assistance is needed to boost agricultural investment, R&D, infrastructure development (particularly irrigation), marketing channels, and farmer finance availability in order to support the agricultural sector's sustainable expansion in Jharkhand (Pandey & Kumari, 2021). Input subsidies are not as productive as public investment when it comes to the overall development of agriculture. Reformulating agricultural policy with reference to public funding of Indian agriculture and reprioritizing state spending is essential. It is noted that while investments are not subject to any upper limit conditions, government of India input subsidies are subject to de minimis conditions under amber box. Reducing the amount of money spent on subsidies and gradually moving toward investment expenditures could be a win-win strategy because they increase farm productivity while also fostering capital formation and capacity building, which increases the likelihood that Indian agriculture will flourish over the long run (Zafar & Tarique, 2023).

Due to intense demand from India's growing population, achieving food security has been increasingly important in recent years. Over time, crop production practices have changed which has led to the emergence of various new issues like crop diversification, problems with growing medicinal plants and biofuel, climate change, an imbalance between the amount and demand of water, the current state of production of high-yielding crop varieties, agricultural crop insurance and pricing, and new developments in globalization and urban encroachment. Globalization has had both beneficial and bad effects on agricultural development through the use of Special Economic Zones (SEZs), and regulations pertaining to globalization must be tightly regulated to lessen the detrimental effects on food security in India (Brahmanand et al., 2013). The goals of food security were initially achieved by the current food gains stocks management system in India, which consists of purchase, buffer stock, and public distribution. However, its flaws and inadequacies have become more apparent, especially in the 1990s and 2000s. To increase food security, the nation must improve its general human development indicators, such as literacy and education.

There have been several instances of stock accumulation of 70–80 million tons since the mid-1990s. The financial costs to the nation are significant. On the other hand, the primary purpose of China's current grain stock programs is to help growers of specific cereal crops in specific locations with regard to price; a "target price" approach is being tested to replace those programs. The liberal trading system that India has seen since the mid-1990s is likewise incompatible with the way food grain stocks are now managed in the nation. To achieve food and nutrition security, India should also support trade liberalization in cereals and noncereal foods. India should, on the other hand, depend less on its high food inventories and more on trade for wheat and rice. Given the rapidly expanding price differential between the domestic and global markets, China might keep importing more food, particularly grain products (Mahendra Dev & Zhong, 2015). India has an abundant supply of food, but as the nation lacks sufficient sources of protein, it must also improve its output of oilseeds, pulses, meat, and meat products. India must use agronomic and management techniques (like watershed development) along with technology (like drought-resistant cultivars) to enhance agricultural and food production. To increase food security, the nation must improve its general human development indicators, such as literacy and education. Despite having lower per capita incomes, certain states, including Kerala and Tamil Nadu, have already attained a better degree of food security. Efforts such as midday meal programs, anganwadi, and employment guarantee programs (MGNREGA), and health programs like Accredited Social Health Activist (ASHA) workers need to be strengthened in order to raise the amount of food entitlements (Reddy, 2016). Achieving self-sufficiency in grain production and establishing the PDS have been the main strengths of the Indian food security system. Famines have been successfully avoided because to the PDS's nationwide distribution of staple foods at reasonable, subsidized prices. India's ability to maintain substantial buffer inventories and its established procurement structure are also among its advantages. The political and administrative will be demonstrated over the years to maintain and advance the PDS. This is seen from the well-implemented policies like the Targeted Public Distribution System (TPDS) and NFSA, among others (Pillay & Kumar, 2018). The right to food is a fundamental right of humans. A less amount of land is being used for agriculture due to urbanization and population growth. Furthermore, the frequency of natural disasters like floods and droughts brought on by climate change brought on by global warming has a negative impact on food supply. Hunger and poverty rise as a result of this. Ensuring food security has emerged as the top priority for global policymakers and government agencies. It will not be enough to rely solely on old methods to ensure food security in light of shifting global dynamics. The interrelationships and impacts of different components are modeled using fuzzy Total Interpretive Structural Modelling (TISM). By decreasing food loss and waste throughout distribution routes, the suggested strategy guarantees sustainability and food security. The primary drivers in the design of an Internet of Things (IoT)-driven food security system are those that promote data collection and recording, such as smart contracts, e-governance and policy development, and the usage of radio frequency identification (RFID) and barcodes. Route optimization using cloud computing, dynamic best before dates via artificial intelligence, and consumption behavior simulation are powered by data from other technologies and may therefore be implemented after the drivers (Kaur, 2019).The COVID-19 pandemic caused an unparalleled worldwide disaster that severely impacted the lives and means of subsistence for millions of laborers. The ability of people to manage their basic needs in the absence of current income is very low in India, where over 90% of the labor force is in the informal sector, and over 80% of working people earn less than Rs. 15,000 a month (according to data from the Periodic Labor Force Survey 2018). The Pradhan Mantri Garib Kalyan Yojana (PMGKY) included a number of food and cash transfer initiatives that were all predicated on payments made to current beneficiaries of different schemes. The PDS offered some assistance, despite the fact that accessing banks and transit limitations have made it difficult in certain cases to get cash transfers. Thousands of migrant laborers were left stranded in various locations, and millions of households were removed off "lists" because they had the necessary paperwork, failed to meet the eligibility requirements, or were denied access because the "quotas" Furthermore, there hasn't been enough focus on other universal initiatives like midday meals and integrated child development services (ICDS) that could help protect the nutrition status of people, especially women and children. The economic downturn and the pandemic's spread persist despite the easing of the lockdown (Sinha, 2021).

Using heteroskedastic linear regression, the variables influencing the degree of urban agriculture participation and its impact on food security in urban cities in Asia and Africa were examined. The impact of the amount of urban agriculture on food security was also examined using the Seemingly Unrelated Regression. The Seemingly Unrelated Regression is a suitable model for this type of analysis because the results of the Breusch–Pagan test for error independence showed that there is a correlation between the determinants of the food-security–access domain tools (i.e., the Household Food Insecurity Access Scale and per capita food expenditure). India's households had moderate food insecurity on average, whereas Ghana's households experienced light food insecurity. The empirical analysis findings showed that different institutional, economic, and demographic factors varied in their influence over the degree to which urban agriculture was practiced and the level of urban food security. Furthermore, food security in Ghana and India was positively impacted by the amount of urban agriculture. The findings also suggest that, in order to lessen urban food insecurity, nations in Asia and Africa should actively encourage urban agriculture among urban households (Bannor et al., 2021). Global access to safe and nutritious food is becoming more difficult as a result of the epidemic, the Russian–Ukrainian War, and climate change. For the poor, especially daily wage workers in some areas, food became increasingly scarce. While vital commodities in India were not subject to travel restrictions during the lockdown, farmers in the nation faced significant challenges in reaching markets, leading to an abundance of food waste. Meanwhile, the middle class's natural hoarding caused the value chain to be disrupted, which made matters worse (NABARD, 2022). India's rural population was impacted by the lockdown. People experienced a reduction in their income as well as adjustments to their food intake and money-generating activities. Rather than fishing, more primary earners in Kerala's coastal regions worked in salaried positions and as daily wage laborers. Phase II saw a decrease in the number of primary earners in Kerala working on their farms but those who did worked longer hours each week. More primary earners began working on their farms outside of Kerala. Every other form of work fell. In phase II, the people with the most working days in a week were those with salaried jobs, those who worked on their farm, and those who were paid on a daily basis in agriculture. The results demonstrate that the lockdown had a detrimental impact on the food and nutritional security of rural households. As the amount of subsidized food grains consumed rose and the amount of more nutritious food items consumed fell, diets became less diversified. The vegetables produced in kitchen gardens are inexpensive and nutritious source of food and generate revenue for rural households if encouraged to plant them (Suresh et al., 2022). Growing climatic risk threatens the food security and means of subsistence of smallholders in vulnerable areas, where farmers frequently lose their entire crop, driving them mostly out of business and into extreme poverty. This highly feminizes agriculture and forces male farmers to migrate elsewhere. South Asia's high rate of rural out-migration is mostly caused by gender and social inequality as well as the region's generally low economic development, particularly in rural communities. Rural communities need to develop economically quickly in order to find a solution. Strategic short-, medium-, and long-term initiatives should be prioritized for business models that create gender-equitable employment opportunities, better education, and training packages. In order to increase women's decision-making and decrease the migration of low-paid workers, a socioeconomic-environmental benefits portfolio should contain targeted, better access and control for women over resources. Adoption of climate-smart agricultural Practices (CSAPs) reduces the demand on government spending in large amounts of compensation paid to farmers as relief during times of climatic risks, therefore promoting gender equality and helping to greatly lessen the impacts of climatic risks (Agarwal et al., 2022). In terms of food security, women in rural and per urban households perform a variety of roles. Not only do adult women experience food insecurity, but their female children are also impacted by unequal gender relations. Mothers not only eat less, less, and last, but their daughters also have this responsibility. Malnutrition over generations and similar illnesses are the result of this. Although women are more likely to contribute to food security in the home and society, they are also more vulnerable to it. Throughout the pandemic, these vulnerabilities were worsened (Dasgupta, 2023).

Increased food security's aspects that are more nutrition-focused are made possible by increasing incomes, which also enable households to buy safer, better-quality, and more varied food. Though direct health and nutrition initiatives are also required to assure the impact, it has been estimated that income development in families and across countries will dramatically reduce malnutrition (Haddad et al., 2003). Individuals' and countries' food security is depending upon demand factors pertaining to wages, risks, health, and social protection measures and domestic caregiving, in addition to supply-side aspects linked to agricultural output. Increased agricultural production frees up manpower and capital for use in other areas of the economy. Industrialization is also aided by agriculture. It supplies markets for industrial products, raw resources for the industrial sector to consume, and exports to generate the foreign exchange required to buy capital goods (Barrett, 2010).

Increasing food production and availability is the primary way that agricultural growth contributes to food security. The world's population gets the majority of its calories from agriculture, which raises cattle and crops for food and raw materials. Food security cannot be ensured solely by the availability of food; it also requires other factors. Poor households can spend more than half of their income on food, and they are frequently net consumers of food in rural areas (Headey, 2011). The degree to which women have control in the home also affects whether or not improved nutrition can be achieved with a larger income. Food, healthcare, and their children's education account for a larger portion of the controllable income for women (Meinzen-Dick et al., 2011).

The challenge of ensuring food security is one that most severely impacts emerging nations with low gross domestic product (GDP) per capita, poor infrastructure, and bad agricultural circumstances. Food security issues are most acute in developing nations where agriculture accounts for a large portion of GDP, where unfavorable environmental conditions impede agricultural output, and where inadequate infrastructure contributes to food security. By adopting farming technologies, improving farmer training programs and extension services, and enacting an open trade policy, countries can increase agricultural productivity and raise the money necessary to finance the import of food to constitute for the shortfall in domestic production. These are the steps that could lead to an oil- or natural resource-based solution to the nutrition problem. In countries facing the challenge of enhancing agricultural productivity, including the most populous ones in the world, policies aimed at improving food security should prioritize the promotion of environmentally friendly technologies, increasing investments in agricultural research and extension systems, and improving farmers' education through technology transfer from developed countries (Pawlak & Kołodziejczak, 2020).

Methodology and Data Source

In this study, production of food grains, nutri coarse cereals, total pulses, and nine oilseeds are measured in million tonnes, and per capita food availability of food grains, nutri coarse cereals, total pulses, and nine oilseeds for the period 2011–2021 are measured in kilograms (kg). All the related data has been collected for the period 2011–2021 from Directorate of Economics and Statistics (DES), an attached office of the Department of Agriculture, Cooperation and Farmers Welfare (DAC & FW), Government of India. This chapter employed descriptive statistics, year over year growth rate and Compound Annual Growth Rate (CAGR).

Analysis of Result

This section presents the analysis of result in two different tables. As per Table 11.1, the highest production is recorded in between 2011 and 2021 for food grains (315.72), followed by the production of nine oil seeds (36.7), total pulses (27.69), and nutri coarse cereals (26.42).

In 2016, the annual growth rate for food grains and nine oilseeds was 9.37 and 23.88%, respectively. In 2019, the highest annual growth rate was 8.31% for nutri coarse cereals. In 2016, the total pulses had the highest annual growth rate of 41.73%. In addition, Table 11.1 shows that the highest compound yearly growth rates are for total pulses and nine oilseeds, respectively, with values of 4.48 and 2.16.

However, it is very evident that the availability per capita has grown from 2011 to 2021 (Table 11.2). The per capita availability of food grains increased from 206.17 kg in 2011 to 224.30 kg in 2021. The per capita availability of total pulses increased from 13.59 kg in 2011 to 19.67 kg in 2021. The per capita availability of nine oil seeds increased from 23.70 kg in 2011 to 26.78 kg in 2021. In contrast, the availability of nutri coarse grains per capita has declined from 21.01 kg in 2011 to 16.09 kg in 2021 (Table 11.2).

Based on the 1996 WFS, food security is defined when all people, at all times, have physical and economic access to sufficient safe and nutritious food that meets their dietary needs and food preferences for an active and healthy life (FAO). Organizations such as the FAO of the United Nations and the WFP often use the metric of "caloric intake per capita" as one of the indicators of food security. So, the "caloric intake per capita" and economic access to sufficient safe and nutritious food can be represented by agricultural production

Table 11.1. Descriptive Statistics, Growth Rate and CAGR of Production (in Million Tonnes) of Food Grains, Nutri Coarse Cereals, Total Pulses, and Nine Oilseeds for the Period 2011–2021.

Crops	Minimum	Maximum	Mean	CV
Food grains	251.54	315.72	277.67	0.08
Nutri coarse cereals	22.15	26.42	24.38	0.05
Total pulses	16.32	27.69	21.36	0.18
Nine oilseeds	25.25	37.7	31.58	0.11

Growth Rate and CAGR of Production of Food Grains, Nutri Coarse Cereals, Total Pulses and Nine Oilseeds for the Period 2011–2021

Year	Food Grains	Nutri Coarse Cereals	Total Pulses	Nine Oilseeds
2011	–	–	–	–
2012	−0.83	−6.28	7.31	3.83
2013	3.08	1.86	4.96	5.85
2014	−4.91	−0.20	−10.91	−16.00
2015	−0.19	−3.10	−4.84	−8.22
2016	9.37	2.54	41.73	23.88
2017	3.60	−2.88	9.90	0.58
2018	0.07	−8.81	−13.14	0.19
2019	4.31	8.31	4.30	5.39
2020	4.45	0.54	10.55	8.22
2021	1.60	−6.09	8.76	4.87
CAGR	1.81	−1.39	4.48	2.16

Source: The author's own calculation.

and per capita availability of the food (Swaminathan & Bhavani, 2013). So, India has managed food security in India successfully in the last decade.

Conclusion and Recommendations

This chapter first measures the annual growth rate and CAGR of food grain, nutri coarse cereal, total pulse, and nine oilseed productions in India from 2011 to 2021. Second, to estimate per capita food grain, nutri coarse cereal, total pulse, and nine oilseed availability (in kg) for 2011–2021 to assess India's food security in the recent decade.

Table 11.2. Per Capita Availability (per Annum) of Food Grains, Nutri Coarse Cereals, Total Pulses, and Nine Oilseeds (in kg) for the Period 2011–2021.

Year	Food Grains	Nutri Coarse Cereals	Total Pulses	Nine Oilseeds
2011	206.17	21.01	13.59	23.70
2012	201.75	19.43	14.39	24.28
2013	205.28	19.53	14.91	25.37
2014	192.79	19.25	13.12	21.04
2015	190.15	18.44	12.34	19.09
2016	205.52	18.68	17.28	23.37
2017	210.46	17.94	18.77	23.23
2018	208.33	16.18	16.13	23.02
2019	215.09	17.34	16.65	24.02
2020	222.53	17.27	18.23	25.75
2021	224.30	16.09	19.67	26.78

Source: The author's own calculation.

The period between 2011 and 2021 had the largest production of food grains, with a reported quantity of 315.72 units. This was followed by the production of nine oil seeds (36.7 units), total pulses (27.69 units), and nutri coarse cereals (26.42 units). The yearly growth rate for food grains in 2016 was 9.37%, while the growth rate for nine oilseeds was 23.88%. Nutri grained cereals saw the greatest yearly growth rate of 8.31% in 2019. In 2016, the whole pulses saw the most significant yearly increase rate of 41.73%. Furthermore, the most significant annual growth rates are seen in the categories of total pulses and nine oilseeds, with respective values of 4.48 and 2.16. Nevertheless, it is abundantly clear that the per capita availability has increased from 2011 to 2021. The per capita food grain availability rose from 206.17 kg in 2011 to 224.30 kg in 2021. The per capita availability of total pulses rose from 13.59 kg in 2011 to 19.67 kg in 2021. The per capita availability of nine oil seeds rose from 23.70 kg in 2011 to 26.78 kg in 2021. Conversely, the per capita availability of nutrient-rich coarse grains has decreased from 21.01 kg in 2011 to 16.09 kg in 2021 which indicates India has managed food security in India successfully in the last decade.

References

Acharya, S. S. (2009). Food security and Indian agriculture: Policies, production performance and marketing environment. *Agricultural Economics Research Review*, *22*(1), 1–19. https://www.indianjournals.com/ijor.aspx?target=ijor:aerr&volume=22&issue=1&article=001

Agarwal, T., Goel, P. A., Gartaula, H., Rai, M., Bijarniya, D., Rahut, D. B., & Jat, M. L. (2022). Gendered impacts of climate-smart agriculture on household food security and labor migration: Insights from Bihar, India. *International Journal of Climate Change Strategies and Management*, *14*(1), 1–19. https://doi.org/10.1108/IJCCSM-01-2020-0004

Bannor, R. K., Sharma, M., & Oppong-Kyeremeh, H. (2021). Extent of urban agriculture and food security: Evidence from Ghana and India. *International Journal of Social Economics*, *48*(3), 437–455. https://doi.org/10.1108/IJSE-08-2020-0519

Barrett, C. B. (2010). Measuring food insecurity. *Science*, *327*(5967), 825–828. https://doi.org/10.1126/science.1182768

Brahmanand, P. S., Kumar, A., Ghosh, S., Chowdhury, S. R., Singandhupe, R. B., Singh, R., Nanda, P., Chakraborthy, H., Srivastava, S. K., & Behera, M. S. (2013). Challenges to food security in India. *104*(7). https://about.jstor.org/terms

Dasgupta, R. (2023). Women, food security and the pandemic in peri-urban and rural West Bengal. *Journal of International Trade Law and Policy*, *22*(3), 209–228. https://doi.org/10.1108/JITLP-04-2023-0020

Food Security in India. (2022). *Drishti IAS*. https://www.drishtiias.com/daily-updates/daily-news-editorials/food-security-in-india

Gautam, M., & Yu, B. (2015). Agricultural productivity growth and drivers: A comparative study of China and India. *China Agricultural Economic Review*, *7*(4), 573–600. https://doi.org/10.1108/CAER-08-2015-0094

Gillespie, S., & Harris, J. (2012). *The agriculture-nutrition disconnect in India what do we know?*. International Food Policy Research Institute, Discussion paper 01187. https://web.archive.org/web/20200319185734id_/http://www.indiaenvironmentportal.org.in/files/file/The%20Agriculture-Nutrition%20Disconnect%20in%20India.pdf

Haddad, L., Alderman, H., Appleton, S., Song, L., & Yohannes, Y. (2003). Reducing child malnutrition: How far does income growth take us? *The World Bank Economic Review*, *17*(1), 107–131. https://doi.org/10.1093/wber/lhg012

Headey, D. (2011). *Was the global food crisis really a crisis? Simulations versus self-reporting*

Kaur, H. (2019). Modelling internet of things driven sustainable food security system. *Benchmarking: An International Journal*, *28*(5), 1740–1760. https://doi.org/10.1108/BIJ-12-2018-0431

Krishi, I. G., & Raipur, V. (2017). *Towards inclusive agricultural development: Growth performance, Welfare challenges and policy innovations*, 79th Annual Conference. Indian Society of Agricultural SciencesCollege of Agriculture.

Mahendra Dev, S., & Zhong, F. (2015). Trade and stock management to achieve national food security in India and China. *China Agricultural Economic Review*, *7*(4), 641–654. https://doi.org/10.1108/CAER-01-2015-0009

Marothia, D., Martin, W., Janaiah, A., Dadhich, C. L., Birthal, P. S., Joshi, P. K., Negi, D. S., & Agarwal, S. (2022). *Re-visiting agricultural policies in the light of globalization experience: The Indian context edited by sources of growth in Indian agriculture: Implications for food security and poverty.*

Meinzen-Dick, R., Behrman, J., Menon, P., & Quisumbing, A. (2011). *Gender: A key dimension linking agricultural programs to in proved nutrition and health.*

NABARD Report. (2022). *Food and nutritional security in India charting the way to a robust agri-food system.*

Pandey, G., & Kumari, S. (2021). Dynamics of agricultural growth and diversification in Eastern India. *Journal of Agribusiness in Developing and Emerging Economies*, *11*(2), 105–120. https://doi.org/10.1108/JADEE-06-2020-0131

Pawlak, K., & Kołodziejczak, M. (2020). The role of agriculture in ensuring food security in developing countries: Considerations in the context of the problem of sustainable food production. *Sustainability*, *12*(13). https://doi.org/10.3390/SU12135488

Pillay, D. P. K., & Kumar, T. K. M. (2018). Food security in India: Evolution, efforts and problems. *Strategic Analysis*, *42*(6), 595–611. https://doi.org/10.1080/09700161.2018.1560916

Reddy, A. A. (2016). Food security indicators in India compared to similar countries. *Current Science*, *111*(4), 632–640. https://doi.org/10.18520/cs/v111/i4/632-640

Saatwik Panigrahi. (2020). A contemporary overview of agricultural productivity: Trends, challenges and lessons for India. *IOSR Journal of Humanities and Social Science (IOSR-JHSS)*, *25*, 1–13. https://doi.org/10.9790/0837-2505070113

Singh, S. (2020). Agriculture development in India: A state level analysis. *South Asian Journal of Social Studies and Economics*, 17–34. https://doi.org/10.9734/sajsse/2020/v6i230162

Sinha, D. (2021). Hunger and food security in the times of Covid-19. *Journal of Social and Economic Development*, *23*, 320–331. https://doi.org/10.1007/s40847-020-00124-y

Suresh, V., Fishman, R., von Lieres, J. S., & Rao, B. R. (2022). Impact of the COVID-19 lockdown on the economic situation and food security of rural households in India. *Journal of Agribusiness in Developing and Emerging Economies*, *12*(3), 491–509. https://doi.org/10.1108/JADEE-07-2021-0177

Swaminathan, M. S., & Bhavani, R. V. (2013). Food production & availability - essential prerequisites for sustainable food security. *Indian Journal of Medical Research*, *138*(3), 383–391.

Swaminathan, M. S., Pandya-Lorch, R., & Yosef, S. (2014). Agriculture and food security. *International Development*, 432–447. https://doi.org/10.1093/ACPROF:OSO/9780199671656.003.0026

Zafar, S., & Tarique, M. (2023). Efficacy of public spending for agricultural development in India: A disaggregate analysis contextualizing subsidies vs investment debate. *International Journal of Social Economics*, *50*(7), 925–940. https://doi.org/10.1108/IJSE-11-2022-0766

Chapter 12

Pakistan: Food Insecurity in a National Security State

Syed Sikander Mehdi

University of Karachi, Pakistan

Abstract

By any count or criterion, Pakistan is a national security state. It always was. The country has fought four wars with India, and it continues to suffer the devastating consequences of confrontation with its formidable neighbor – India; it has been put under siege for more than 40 years by the terrorist and religious extremist groups; it has suffered long spells of military rule; its political structure remains fragile; its civil society organizations are vulnerable, weak, and toothless; and its media, universities, research institutes, and judiciary are under huge stress. Given the state of affairs, the realization is rather slow to grow that human development strengthens national power and national security much more than weapons. It is in this context that a study on food insecurity in Pakistan has been undertaken. It aims to explain as to how a predominantly agricultural country has been reduced into a food resource poor country, and how has the neglect of social, economic, and cultural sectors made Pakistan more insecure. This study covers a number of related issues as well, but it mainly looks for answer to two questions: first, should food insecurity be studied in isolation as a separate issue or as a part of the larger issue of governance, security, and people; and second, isn't there a need to drastically restructure the economic, political, and security landscape and framework to significantly diminish poverty, illiteracy, and food insecurity in the country?

Keywords: Imagination; freedom; hunger; international food system; feudalism; food security

Growth, Poverty and Developmental Aspects of Agriculture, 177–190

Published under exclusive licence by Emerald Publishing Limited
doi:10.1108/978-1-83608-076-320241012

Introduction

Shorn of all its embellishments, a national security state is essentially a state where the military is the king. The military dominates and often dictates other institutions of governance like the executive, legislature, and judiciary and controls the media, universities, and civil society organizations. It may rule directly or it may rule indirectly through the façade of an elected civilian government. The main thrust, direction, and focus of its policies and future planning remain unchanged. Serving the best interests of the military establishment, keeping the forces and their families satisfied and happy, crushing the movements for democracy, human security, and justice, ensuring that it would never be brought to book by the parliament, civilian government, or judiciary for its high-handedness and unconstitutional practices, and safeguarding the citadel of power at all costs are usually the dominant concerns behind its policies pursued in the name of national security. Furthermore, a national security state benefits immensely by hosting a permanent enemy in the neighborhood or on a faraway territory, keeping the nation agitated in the name of nationalism, patriotism, ideology, and security, and exercising vigilance to oversee that the new political, economic, cultural, and security ideas, new knowledge, and visions of new pasts and new futures don't gain ground.

Furthermore, a national security state thrives on fear and insecurity it plants all around and instils fear in the hearts and minds of the citizens it is supposed to serve. The national security state may put its proxies on the stage to run the country. However, the state is run for all practical purposes, by not so visible executive comprising serving and retired military officials and their cronies. With gun in hand, they project themselves as the most competent administrators, clever decision-makers, brilliant short-term and long-term planners and executors, and brave risk takers. Being trained to look at the killing and wounding of unarmed and innocent men, women, and children caught in wars, conflicts and violence, and destruction of properties and infrastructure as collateral damage, a national security state isn't bothered much by human tragedies – often produced by its own policies. Likewise, it is seldom uncomfortable with its own dehumanizing approach to governance. It enjoys power, which grows out of the barrel of the gun and because it believes in the dictum that the end justifies the means. Drunk with power and driven by a sort of religious zeal to crush any opposition to its rule and its way of governance, it exercises all the power at its disposal to disempower the society and the people and reassure the power elites that their interests will be served at all costs. Such a reassurance is sought all the time by the accomplices and beneficiaries of a national security state – the military, the feudal, religious, bureaucratic, judicial, political, business, and media elites as well as the self-serving intellectual elite. It forces the people to live forever on the margin and beyond the margin in the valley of the unheard, unseen, and unmourned. The rule of military generals in Pakistan like the rule of Ayub Khan (October 1958–March 1969), Yahya Khan (March 1969–December 1971), Zia-ul-Huq (July 1977–August 1988), and Pervez Musharraf (October 1999–August 2008) clearly indicates their little respect for the constitution and for the common citizens and their longing for democratic politics and culture.

The issue of food insecurity is rarely taken up in its entirety. Likewise, a long-term approach is seldom taken. Again, the national security state, which is inherently a dehumanizing state, deals with its starving population as a number and that's all. The state may or may not feel bad about the people dying of hunger, but it feels bad about the bad media coverage at home and abroad. The realization is usually minimal that the starving humans – the individuals, families, and communities – are victims of preventable man-made and system-made tragedies. Similarly, the developed and resourceful countries extending aid and assistance to the hungry fail to realize that human indignities and deprivations like poverty, inequality, illiteracy, slavery of all kinds, and hunger and food insecurity are man-made and system-made tragedies. As such, they show little interest in going to the roots to wipe out the curses. Sooner or later, donor fatigue surfaces and food aid is reduced or stopped by some reluctant donors. Soon after starts the bashing of international food support system by the food aid addicted countries including Pakistan. As a matter of fact, and being a kind of national security state which is obsessed with hard security concerns, food insecurity may fail to draw their attention toward itself

Since history and silence and fear and unfreedom move together in Pakistan, it is rather problematic to convince the policymakers that it is unfreedom that causes insecurity and decline. Small wonder therefore that the crucial decisions of the country – taken from the early period of post-colonial Pakistan onward – were taken without much debate and discussion. Again, the decisions were always defended in the name of security of the state and people. With the passage of time, the areas of national concern were enlarged for the power elites, and military involvement in the political, economic, and social sectors increased. As a result, security of political, economic, and social system – so cleverly crafted by the power elites and so well entrenched now – has become a permanent part of the larger national security agenda. Any attempt to bring about radical policy shift, initiate structural changes, and reprioritize the major concerns of the citizens is resisted by the state (Ahmad, 1989; Ali, 2014; Khawaja, 2016; Niaz, 2010; Siddiqui, 2015). The issue of food insecurity becomes all the more problematic because it is commonly viewed as a number issue and not as a preventable human catastrophe, causing heavy blows to the efforts to build a global human order. Some of the points raised here have been explained below to present the issue of hunger and food insecurity in a wider context and in the context Pakistan.

Imagination, Freedom, and Food Insecurity

The national security state of Pakistan, like other states and other kinds of states, lives and breathes in an age of specialization and enjoys the fruits of specialization. Facing the challenge of food insecurity, it responds as so many other states respond, and it is usually a lukewarm response. It couldn't have been otherwise because we live in an age of power politics and electoral politics. Constituting an insignificant portion of the population of any country and being destitute and powerless, the hungry and starving humans don't matter much. In the case of a national security state, they matter much less. A dilemma therefore sneaks in. On the one hand, there is clearly an emergency to reach out to the people dying and suffering because of hunger and food insecurity, and on the other, there is no emergency because these starving people are usually incapable of rising for their rights and freedom and kicking up a political storm. Worse still, the dependent economies like Pakistan's economy await eagerly the arrival of food aid and assistance from different international sources. Sometimes, the foreign experts or native specialists, preferably foreign educated, are approached to recommend quick damage control measures. Sometimes, a state like the Pakistani state registers its seriousness by taking measures like proper watering, ploughing, and seeding of the agrarian lands, fighting the pesticides and other agents of soil erosion and desertification, protecting crop loses, increasing the yields by leapfrogging toward mechanization and modernization of agriculture, and offering token incentives to the farmers. These are important steps and need to be taken. However, such steps and several other related steps are just routine measures. There is nothing spectacular or innovative about them, and they cannot produce solid and trailblazing results in the short-term and even in the longer term. As such, imagination will have to be summoned to strategize for a better and humanistic agrarian future. The problem with Pakistan and many other countries is that they try to reduce food insecurity by isolating the issue from a chain of other connected issues, and such an approach doesn't take the countries far enough. In a way, the real problem is in looking for the tree and missing the forest.

The Challenge

According to a peer-reviewed report on hunger prepared in collaboration between German nonprofit Welthungerhilfe and its Irish counterpart Concern Worldwide released in October 2023 (latest edition), Pakistan ranked 102nd out of 125 countries, and its Global Hunger Index (GHI) score of 26.6 placed it at a level of hunger that is *serious* (www.globalhungerindex.org). A year before, it was ranked 99th out of 126 countries, and its GHI score had dropped from 38.1 in 2006 to 26.1 in 2022. However, this positive trend couldn't be maintained in the following year, and a quick improvement in the near future seems very unlikely. Various international reports including UN reports paint a bleak future for three most affected provinces: Baluchistan, Khyber Pakhtunkhwa, and Sindh. A report, prepared jointly by the Food and Agriculture Organization (FAO) and World Food Programme (WFP), released in October 2023, identifies 18 "hunger hotspots" in a total of 22 countries where food security was expected to significantly deteriorate and offered the outlook for the next 6 months from November 2023 to April 2024. It places Pakistan along with Afghanistan, Democratic Republic of Congo, Ethiopia, Haiti, Somalia, Syrian Arab Republic, and Yemen in the category of countries needing urgent action to avert extreme hunger or death. It identifies these countries as "hotpots of very high concern." It informs that 2.2 million people were projected to face emergency level of acute food insecurity (IPC Phase 4) between November 2023 and January 2024, including two areas in Khyber Pakhtunkhwa, where 20% of the population was projected to face critical levels of acute food insecurity (IPC Phase 4) (FAO and WFP) (www.sbp.org.pk).

Pakistan's precarious food insecurity situation can be gauged from the fact that almost half of its children under 5 are stunted (low height for age), and 1 in 10 suffers from wasting (low weight for height). Again, Pakistan is among those seven countries that cumulatively account for two-thirds of world's undernourished populations along with Bangladesh, China, Congo, Ethiopia, India, and Indonesia (www.sbp.org.pk). It may also be noted that Pakistan is placed in South Asia – a region which international development practitioner and researcher Sachi Satapathy describes as "world's epicenter of food insecurity," where roughly 1.4 billion people are "unable to afford even the cheapest, locally available healthy meal," and where 55.5% of Sri Lankans, 66.1% of Bangladeshis, 74.1% of Indians, 76.4% of Nepalese, and 82.8% of Pakistanis faced serious difficulties in arranging healthy food for themselves and their families during 2021–2022 fiscal year (www.sbp.org.pk).

One can have some idea from the above details regarding the magnitude of the challenge Pakistan faces in tackling hunger and food insecurity issues in the country. However, the situation would look much worse if viewed in the wider context. First, the very fact that Pakistan's food insecurity is more severe than the food insecurity being faced by 101 countries above it in the rank is worrisome. Again, this ranking doesn't reveal enough because the issue of food insecurity is not only about hunger and starvation. Broadly speaking, it is about a host of inter-related, visible, and invisible challenges. It is, for example, about deprivations of all sorts and about their devastating consequences; it is about man-made and system-made human tragedies, which can be averted; it is about the silent and unnoticed genocides taking place in parts of the world because of hunger and malnutrition and about filling in of the graveyards; it is about the issue of food scarcity in the midst of plenty; it is about the unwillingness of the stakeholders involved in global governance – especially the rich, developed, and resourceful countries – to acknowledge their responsibility toward the entire planetary population; and it is about realizing that even the survivors aren't spared by the monster called hunger as atomic survivors were not spared by the radiation effects of the atomic bombs dropped on Hiroshima and Nagasaki on August 6 and 9, 1945. The hunger survivors, especially the children, suffer physical and mental incapacities and disabilities because of retardation or slow physical and intellectual growth caused by irregular and insufficient food intake, malnutrition, and so on. Consequently, hundreds of thousands of hunger survivors suffer the deprivation of the right to growth, education, good health, happiness, hope, and freedom.

The second point relates to the linkage between human development, economy, and security (Najam, 2005). Since Pakistan is a national security state, it is deeply concerned about the hard security issues or battlefield security issues. As such, its national security discourse as well as its national economic discourse generally doesn't highlight the importance of nonmilitary sources of security like powerful, free, and energetic parliament, independent judiciary, free media, dynamic civil society organizations, colleges and universities as centers of new ideas and movements, transparent electoral politics, women's empowerment and healthy, educated, skilled, and motivated citizens in all walks of life. Also little highlighted is the cost of hunger and malnutrition to national economy. According to a report of State Bank of Pakistan, under-5 malnutrition costs around US $ 7.5 billion every year to the country, which is equivalent to 3% of its gross domestic product (GDP) (www.sbp.org.pk).

The third point which deserves mention is the fact that Pakistan is endowed with remarkable agricultural resources and potentials. However, now it is facing the challenge of food shortage for a variety of reasons (Shaheen, 2005). It may, in fact, experience far more severe food insecurity in the future because of its growing inability to produce enough food stuff for its phenomenal and rapid increasing population and also due to food smuggling and trafficking into neighboring Afghanistan. The failure is more deplorable because Article 38 of Pakistan's constitution says that the state shall provide basic necessities of life like food, clothing, housing, education, and medical relief to its citizens (Sadeq, 2009). Nevertheless, government sources continue to project the country as self-sufficient in major staples: ranked at 8th in producing wheat in the world, 10th in rice, 5th in sugarcane, and 4th in milk production (Sadeq, 2009). One may add here that the agricultural sector, which was the mainstay of Pakistan's economy in the past, will remain very important for the economy and the people for a long time to come. The agriculture sector is very important because its contribution is estimated to be about 22.7 % of its GDP. Moreover, it employs 37.4% of national labor force, and about 70% of Pakistan's exports are directly or indirectly derived from agriculture. Again, agriculture covers an area of 30.5 million hectares, and the agricultural land constitutes 47% of the national land, which is higher than the global average of 38% (FAO, 2024).

However, even when Pakistan was known as a major food-producing country in South Asia, part of the country and part of its people were experiencing acute hunger and food insecurity because of market-based food availability system and because the poorest segment of the population didn't have the purchasing power nor easy access to food. With the passage of time, the country ceased to produce enough for its entire population. Subsequently, the number of poverty-stricken and malnutrition-stricken people increased steadily. A number of reasons, mostly technical and economic, are usually put forward by government officials, baptized experts including economists, and controlled media, civil society organizations, universities, and research centers to explain the depletion of food resources and self-sufficiency. The reasons put forward include a high rate of population growth, inadequate incentives given to farmers, slow mechanization of agriculture, faulty system of water distribution for the agricultural lands, frequent arrival of floods, and other impacts of climate change, COVID-19 pandemic, lack of long-term planning, rapidly dwindling financial capacity of the country to invest substantially in the agriculture sector, soil erosion, lack of long-term agricultural planning, and ever spreading corruption and terrorism. However, certain factors and issues, which may not seem to be very directly related to the issue of food insecurity in Pakistan and which are rarely discussed, do significantly contribute to the spread and entrenchment of poverty and food insecurity. Unless these are taken care of, the chances of marked improvement in any area of national concern will remain problematic. Some of these are: hold of landed aristocracy in the rural and semi-urban areas, powerlessness of the masses including the farmers, authoritarian political culture, paramountcy of the military elite in national economic, political, and security policymaking and decision-making process (Anatol, 2011; Aqil, 2014; Pasha, 2008; Siddiqa, 2007; Ayaz, 2013), cost of confrontation with India for over 70 years, and deteriorating relations with two other bordering countries – Afghanistan and Iran – least investment in education, health, and other sectors of human development, enslavement of politics and governance, and conversion of the country into a vast thought desert. These and other "gridlocks will have to be broken" (UNDP, 2023/2024) if Pakistan wishes to move forward and eliminate poverty, food insecurity, and illiteracy.

At the moment, Pakistan is in the throes of "policy crisis," a term which refers to "major economic and social crises that include severe effects of climate change, impacts from the COVID-19 pandemic, and the effects of Russian invasion of Ukraine – has caused severe disruptions to energy, food, and finance system" in the Asian and Pacific region (UNDP, 2024). As a matter of fact, the continuing war in Ukraine and the massacre of the Palestinians in Gaza by Israel followed by skirmishes between Iran and Israel posing grave threat to regional security are also disturbing the international trade chain including the food supply chain and causing enormous increase in energy prices and transportation cost. In the case of Pakistan, the policy crisis also includes serious issues of governance, flagrant imbalance in the share of power, responsibilities, and opportunities being availed by the main pillars of the national security state, mushrooming unfreedoms, and powerlessness of the masses. However, the country doesn't seem to be willing to go for a big leap forward. As such, it seems more likely that it will continue to respond to the challenge of poverty, hunger, and food insecurity as it had done in the past: taking symbolic, demonstrable, and rather small and half-hearted measures to improve the agricultural sector and diminish the level of food insecurity. Since the national security state is not in a mood to abdicate the throne and Pakistan finds itself in one of the gravest economic challenges it had ever faced (Mehdi, Forthcoming), it is more likely that the country will look toward the international food system and its affiliated organizations for larger food aid and assistance. Now the big question is this: can the international food systems bail out Pakistan?

International Food System and Pakistan

Literature on food system abound. In fact, hundreds, perhaps thousands, of research studies and reports have been prepared during the last sixty years or so by the UN and its specialized agencies, the UN member states, independent research centres, civil society bodies and donor agencies. One such study produced recently by the Organization for Economic Cooperation and Development is *Understanding the Global Food System* (OECD, 2024). These highlight the extent and limit of the role of international food system in dealing with the multi-faceted challenges posed by hunger, malnutrition and food insecurity in different parts of the world. In fact, poverty, illiteracy, hunger and food insecurity reduction and elimination have all along been one of the major concerns of the UN. The Food and Agriculture Organization, one may point out here, was the first permanent specialized agency to be established by the UN in October 1945, the very month and year when the UN itself was established. While UN's Universal Declaration of Human Rights, a historic document proclaimed by the UN General Assembly in Paris on December 10, 1948 does not specifically mention the issue of hunger and food insecurity, a number of its articles like Article 3 ("Everyone has the right to life, liberty and security of the person"), Article 4 ("No one shall be held in slavery or servitude, slavery and the slave trade shall be prohibited in all their forms") and Article 22 ("Everyone, as a member of society, has the right to social security and is entitled to realization, through national effort and international co-operation and in accordance with the organization and resources of each State, of the economic, social and cultural rights indispensable for his dignity and the free development of his personality") can be related to the destitute and hungry in any country. In the year 1960, the Freedom from Hunger Campaign was launched by FAO to eradicate hunger once and for all. During the period stretching from 1960 to 1992, strong campaigns were launched within the UN and outside to establish a new economic world order and diminish the nontraditional sources of insecurity, including food insecurity.

A major initiative in this regard was taken by the UN in September 2000, when leaders of 189 member states signed the historic Millennium Declaration and committed themselves to achieve eight measurable development goals including the goal to eliminate extreme poverty and hunger by 2015 (MDGs 2000–2015). While some progress was made during 2000–2015 (Jackson, 2007), this global agenda was replaced by a more ambitious future plan. Commonly known as Sustainable Development Goals (SDGs – 2030), this enlarged global agenda set 17 goals for the world to achieve. The second SDGs goal was to "End hunger, achieve food security and improved nutrition and promote sustainable agriculture" by 2030 (UN, 2024). Along with the UN System, a number of international research institutes and donor organizations have been actively involved in focusing on various aspects of hunger and food insecurity issues.

Action against Hunger is an international humanitarian organization. More than 40 years ago, in 1979, it was established by a group of French doctors, scientists, and writers to campaign to end hunger. In the year 2022, it worked in 56 countries around the world with more than 8,990 employees helping 28 million people. One of its reports published in August 2023 informs that: (a) 3.1 billion people were unable to afford a healthy diet; (b) 783 million people suffered from hunger around the world; (c) 1 in 10 people suffered from hunger; (d) 80 million more women than men were hungry in 2022; (e) 85% of people facing hunger lived in conflict-affected countries; (f) 13.6 million children suffered from severe malnutrition; (g) 2 million children died every year from malnutrition; (h) 45% of all child deaths worldwide were from hunger and other related causes; and (i) 1 in 3 children in low and middle income countries suffered from chronic malnutrition (www.globalhungerindex.org). Another report of Action Against Hunger, *Hunger Funding Gap Report*, says that global hunger funding gap had hit 65% in 2023 for the countries with most urgent needs (HGP, 2024). As if all this wasn't tragic enough, the *UNEP Food Waste Index Report 2024* adds salt in the wounds. It informs that 1.5 billion tonnes of food waste (including inedible parts) is generated per year, totaling 79 kilos of food waste per person (Butler, 2024).

The state of affairs being what it is, it is unlikely that the aid, assistance, and support provided by the global food system can significantly reduce Pakistan's food woes and worries. Nevertheless, it would be wrong to say that the country didn't benefit from the international network of food aid and assistance. In 2022, for instance, WFP – food arm of the UN System and largest humanitarian organization established in 1961 – provided relief, recovery, and development assistance to about 4.8 million people – more than five times the planned number. Large-scale devastating floods precipitated by monsoon rains had produced the surge in the number of people needing help and assistance on an emergency basis. The floods had affected 33 million people – killing over 1,700 people, displacing 8 million, and causing USD 30 billion in damage and losses. Hardest hit were the food, livestock, and fisheries sectors. WFP reports that two-thirds of those it had supported during 2022 – 3.1 million people – suffered emergency levels of food insecurity and needed life – saving help (WFP, 2022). In addition, a number of national and international nongovernmental organizations (NGOs) and Charities, including the Eidhi Foundation, Food Savers, Action Against Hunger, Global Alliance for Improved Nutrition (GAIN), Nutritional International, have been involved in various poverty and food-related programs in Pakistan. However, a few umbrellas doled by some individuals and organizations cannot be enough to face the heavy monsoon rains.

Keeping in view the shortcomings, failures and limited role of donation-based global food system, predominantly underfunded and corruption – and nepotism – loaded local and national food programs, and distribution of free meals on the streets by some philanthropic organizations in Pakistan, it can be said with some confidence that hunger and food insecurity in the country cannot be eliminated through such measures. Such measures are like removing small stones on the road to embark on a journey of a 1000 miles and not realizing that the road has been blocked from all sides not by the little stones but by the huge and ugly obstacles. Unless these obstacles are removed, the road to the future will remain blocked. However, there are obstacles which are rarely mentioned and discussed even in the serious studies on Pakistan's economy. Admittedly, a number of good studies produced by prominent Pakistani economists do discuss the agriculture sector and its ailments in great detail and also offer recommendations for early recovery and improvement. However, they rarely discuss Pakistan's defence expenditure and its unaccountability by the parliament, the cost and consequences of confrontation with India, the cost of deteriorating relations with two other bordering countries – Afghanistan and Iran, and management of politics, including electoral politics and formation and dissolution of governments by the military (Husain, 2018, 2019; Pasha, 2019; Zaidi, 2022). Of course, some studies do cover such issues (Jalal, 1991; Nawaz, 2008; Rahman, 2022; Shah, 2014; Siddiqa, 2007). A recent study argues that Pakistan's economy would be on the right track if all the institutions of governance remain within the limits specified by the constitution (Hussain, 2023). In other words, a clarity in thinking seems to be emerging that Pakistan's economic woes and worries including the food insecurity issue need to be viewed in a broader context. In addition, the severe effects of war on economy on national exchequer can no longer be kept out of political and economic discourse because everything is relational.

Everything Is Relational

The wounds of colonialism fester for long, very long. In the case of postcolonial Pakistan, as in the case of several other countries, the wounds have remained largely unhealed. Now, the power, resources, and opportunities are being enjoyed by *native colonial lords*. Like their former colonial masters, they also rule with iron fists, abhor peace and change, and nurture contempt for the common people. Clearly, such a society wouldn't be duty bound to eliminate hunger and malnourishment in the country in the shortest possible period. How else can this attitude be described if not as a huge obstacle to the journey toward freedom, human development, and democratic governance? The continuance of feudal practices in the rural and semi-urban areas of Pakistan is another big obstacle to Pakistan's transformation into a more just, equal, and progressive society.

It is, indeed, a historical truism that the abolishment of feudalism in all its forms and manifestations lets loose the forces of change, helps the people and the country to leap out of decadent traditions and practices, and leads the society toward the world of new knowledge and freedom, political wisdom and stability, and development for all. Several eminent progressive writers of Pakistan have written extensively on the devastating consequences of feudalism on its social and political culture, governance, and value system (Ahmad, 2003; Ali, 1996; Alvi, 2000, 2002, 2014). However, before we proceed further, it is desirable to ask a question, which seems to be very pertinent in the context of food insecurity in Pakistan. The question is this: is feudalism already a memory in the country or does it still hound around? Some Pakistani scholars, including the distinguished economist Akbar Zaidi, currently serving as Director of a leading educational institution of the country – Institute of Business Administration (IBA) – are of the view that feudalism is no more or not so pervasive, active, and influential as it used to be for centuries. Writing a full chapter on popular perceptions and decline of feudalism in Pakistan and another full chapter on green revolution and land reforms in his phenomenal study on Pakistan's economy (a study which has run into several editions), Zaidi observes that "feudalism despite popular perception, does not dominate production in the agricultural sector" and that while "feudal enclaves" existed at the time of independence, their trend was on the decline, being replaced by an altogether different and new order. Another eminent economist and widely respected public intellectual Kaiser Bengali often talks about the continuing influence of feudalism on power politics and national economy in his lectures at different forums.

Feudalism is like an octopus. It has many tentacles, and it swings them in different direction. As an oppressive and suppressive political and economic system, it was functional in the Pakistani territories even before the British had arrived. A study on landlord power in Sindh during 1865–1901, based on author's PhD thesis, discusses as to how were the Sindhi waderas given the power to treat the lands under their occupation as fiefdoms and its people as serfs in return for their loyalty and support to colonial order. He concludes by saying that "for Sind, the *Waderos* were everything; the British were nothing. It was the *Waderos* who effectively governed Sind, as they still do (Cheesman, 1997: 225)." The British have left, but the arrangement in a new form remains intact and in practice in the postcolonial Pakistan. The only difference one may discern is a more visible and active role of the rapidly modernizing landed aristocracy in local and national politics (Jafri, 2011; Khurshid, 2014). Small wonder therefore that alternate local political leadership, a leadership strong and daring enough, to challenge the power of the big feudal lords in different political and electoral constituencies in rural Pakistan is still awaited. This gridlock needs to be broken so that human issues like poverty, illiteracy, food insecurity, and powerlessness of the masses are dealt with more directly and effectively.

Uncontrolled population growth is another big obstacle which needs to be seriously attended to on an urgent basis, if economic, social, and political insecurity, including food insecurity, are targeted to be diminished to a significant extent. Likewise, the obsession with growth only needs to be controlled, and the emerging concepts like womanomics, human economics, and freedom and peace economics need to be introduced into mainstream discourse on economy, politics, and culture and their relevance examined passionately. In the same way, the national security state needs to be transformed into a human welfare and human security society. Then and only then, the human issues relating to health, poverty, illiteracy, and food insecurity can be attended to on a permanent basis.

Conclusion

Common sense says that the future of Pakistan depends on the future of food security in Pakistan. If the country fails to tackle the economic challenge fast enough, it would go deeper and deeper into the quagmire. As a result, poverty, hunger, and food insecurity would multiply. The challenge the country has been facing is really huge. As media reports and statements of officials indicate, collapsing economy is an existential issue for Pakistan. As such, the country will have to focus, with all the resources and opportunities at its disposal, on rapid economic recovery in order to stride ahead with full confidence. All this seems very unlikely to be achieved in the near and even in not so near future. Let us assume for argument sake that Pakistan's economy does turn around, but what is the guarantee that the country would then devote its energy and resources to human development programs and projects, and make all out efforts to uplift 40%

of its population now living below the poverty line. These are the homeless, sick, vulnerable and starving people. As such, mere development is not enough, development for whom is important. Again, the entrenchment of fascism in the name of development won't diminish the woes, worries, and pains of the common people, including hundreds of thousands of hungry and undernourished citizens. Pakistan can be saved if it realizes that it has been destroyed because of authoritarian rule (Akhtar, 2022; Jan, 2021), frequent derailment of democracy (Hoodbhoy, 2021), military interference in civilian matters, never-ending terrorism and religious extremism, and powerlessness of the masses. Sometimes, development without freedom can be achieved faster as it was achieved in the Soviet Union under Stalin or in Germany under Hitler, but the gross neglect of the poor, hungry, undernourished, and powerless often results in anarchy, outbreak of group and crowd violence, and collapse of the national security state itself.

Pakistan stands today at the edge of the precipice. However, the national security state may still be unwilling to accept that the country is paying the price today for the ruthless bulldozing of the movements for freedom, democracy, and human rights launched in the country. The movements include those launched by the students, poets, writers, teachers, journalists, women, lawyers, and farmers against authoritarian regimes. Among these, the Movement for the Restoration of Democracy boldly challenged the draconian regime of General Zi ul Haq in the 1980s. Under the leadership of former student leader Fatahyab Ali Khan, the *Mazdoor Kisan Party* (Labor and Farmers' Party) played a significant role in shaking the very foundation of Zia's dictatorial regime (Hasan, 2022). The memory of such movements needs to be kept alive and new movements launched. Building peace within and building peace across the borders and beyond, launching powerful movements for freedom, equal rights, human development based on human economics, and drastic curtailment of military's role in politics, economics, and business may help transform the national security state of Pakistan into a human society, where concerted and sincere efforts can be made to significantly diminish the pains and pangs of poverty, hunger, and malnutrition. There is no other way to a better future for all.

References

Ahmad, S. S. (1989). *Struggle against Pakistan*. Justice Kayani Memorial Law Society.
Ahmad, F. (2003). *Imperialism and Pakistan: 1947 to 2002* (in Urdu). Takhleeqat.
Akhtar, A. S. (2022). *The struggle for hegemony in Pakistan*. Folio Books.
Ali, M. (1996). *Feudalism* (in Urdu). Fiction House.
Ali, M. (2014). *History of martial law in Pakistan* (in Urdu). Tareekh Publications.
Alvi, H. (2000). *Feudalism and imperialism* (in Urdu). Fiction House.
Alvi, H. (2002). *Pakistan: State and its crisis* (in Urdu). Fiction House.
Alvi, H. (2014). *Pakistani aristocracy, military and bureaucracy* (in Urdu). Tareekh Publications.
Anatol, L. (2011). *Pakistan: A hard country*. The Penguin Group.

Aqil, S. (2014). *The army and democracy: military politics in Pakistan*. Harvard University Press.

Ayaz, B. (2013). *What's wrong with Pakistan?*. Hay House India.

Cheesman, D. (1997). *Landlord power and rural indebtedness in colonial Sind: 1865–1901*. Curzon Press.

Hasan, M. (2022). *Pakistan in an age of turbulence*. Pen & Sword History.

Hoodbhoy, N. (2021). *Aboard the democracy train: Pakistan tracks the threat within*. Paramount Books (Pvt.) Ltd.

Husain, I. (2018). *Governing the ungovernable: Institutional reforms for democratic governance*. Oxford University Press.

Husain, I. (2019). *Pakistan: The economy of an elitist state*. Oxford University Press.

Hussain, A. (2023). *Pakistan, institutional instability & underdevelopment: State, people and consciousness*. Folio Books.

Jafri, A. A. (2011). *Pakistan's big political landlords* (in Urdu). Virsa Publications.

Jalal, A. (1991). *The state of martial law: The origins of Pakistan's political economy of defence*. Vanguard Books Pvt. Ltd.

Jan, A. A. (2021). *Rule by fear: Eight theses on authoritarianism in Pakistan*. Folio Books.

Khawaja, A. (2016). *People's movements in Pakistan*. Kitab Publishers.

Khurshid, M. (2014). *Feudal ascendancy and the role of Sir Fazl-e-Husain in the Punjab politics (1901 1936)*. National Book Foundation.

Mehdi, S. S. (Forthcoming). *Development as freedom and Pakistan's handcuffed economy*.

Najam, A. (2005). Broadening the security debate: Human and environmental dimensions. In *Sustainable development: Bridging the research/policy gaps in southern contexts* (Vol. 1, pp. 31–51). Oxford University Press.

Nawaz, S. (2008). *Crossed swords: Pakistan, its army, and the wars within*. Oxford University Press.

Niaz, I. (2010). *The culture of power and governance of Pakistan: 1947–2008*. Oxford University Press.

Pasha, H. A. (2019). *Growth and inequality in Pakistan: Agenda for reforms*. Friedrich-Ebert-Stiftung.

Pasha, S. (2008). *Crossed swords: Pakistan, its army, and the wars within*. Oxford University Press.

Rahman, T. (2022). *Pakistan's wars: An alternative history*. Folio Books.

Sadeq, N. (2009). Pakistan's food security: Unnecessary risks and potential prospects. In J. Bennet (Ed.), *Pakistan: Haunting shadows of human security* (pp. 163–201). Bangladesh Institute of International and Strategic Studies.

Shah, A. (2014). *The army and democracy: Military politics in Pakistan*. Harvard University Press.

Shaheen, F. H. (2005). Agriculture – From public good to private asset – Consequences of policy inaction in rural Pakistan. In *Sustainable development: Bridging the research/policy gaps in southern context* (Vol. 1, pp. 77–88). Oxford University Press.

Siddiqa, A. (2007). *The military Inc.: Inside Pakistan military economy*. Pluto Press.

Siddiqui, M. A. (2015). *Pakistan: From religion to fascism*. Ushba Publishing International.

Zaidi, S. A. (2022). *Issues in Pakistan's economy*. Oxford University Press.

Online Sources

Action Against Hunger. *10 Facts about world hunger.* https://mcusercontent.com/a95cced314f76950628adc472/files/a9806e33-0608-e78b-75e1-63ebc1630c2c/Action_Against_Hunger_10_Facts_About_World_Hunger.pdf. Accessed on April 10, 2024.

Butler, S. (2024, March 28). UN report reveals colossal levels of food waste. *Geographical.* https://geographical.co.uk/news/un-report-reveals-colossal-levels-of-food-waste Accessed on April 18, 2024.

FAO. https://www.fao.org/pakistan/our-office/pakistan-at-a-glance/en/. Accessed on April 12, 2024.

Global Hunger Index-Pakistan. https://www.globalhungerindex.org/pakistan.html. Accessed on April 14, 2024.

HGP. (2024). *Hunger funding gap report.* https://reliefweb.int/report/afghanistan/action-against-hunger-2024-hunger-funding-gap-report. Accessed on April 10, 2024.

Hunger Hotspots: FAO-WFP warnings on acute food insecurity, November 2023 to April 2024 outlook. https://docs.wfp.org/api/documents/WFP-0000153539/download/?_ga=2.209391954.111913743.1713278484-1764282634.1709972459. Accessed on April 16, 2024.

Jackson, P. (2007, December). A prehistory of the Millennium Development Goals: Four decades of struggle for development in the United Nations. *United Nations, UN Chronicle, 44*(4), 6-7. https://www.un.org/en/chronicle/article/prehistory-millennium-development-goals-four-decades-struggle-development-united-nations. Accessed on April 8, 2024.

OECD. (2024). https://www.oecd.org/agriculture/understanding-the-global-food-system/. Accessed on April 18, 2024.

State Bank of Pakistan. *Special Section-2: The state of food security in Pakistan.* https: www.sbp.org.pk. Accessed on April 18, 2024.

UN. (2024). *United Nations, peace, dignity and equality on a healthy planet – Global issues: Food.* https://www.un.org/en/global-issues/food. Accessed on April 7, 2024.

UNDP. (2023/2024). *This expression has been taken from UNDP Human Development Report 2023/2024: Breaking the gridlock: Reimagining cooperation in a polarized world.* https://reliefweb.int/report/world/human-development-report-20232024-breaking-gridlock-reimagining-cooperation-polarized-world-enarruzhjp. Accessed on April 21, 2024.

UNDP. (2024). The term 'polycrisis' has been derived from a report prepared by the UNDP, Economic and Social Commission for Asia and the Pacific (ESCAP) and Asian Development Bank. *Delivering on the sustainable development of goals through solutions at the energy, food and finance nexus.* https://www.undp.org/asia-pacific/press-releases/global-crises-further-threaten-progress-sdgs-adb-un-report. Accessed on April 17, 2024.

WFP Pakistan: 2022 Annual Country Summary. https://docs.wfp.org/api/documents/WFP-0000151578/download/?_ga=2.65418606.199083949.1713457269-1764282634.1709972459. Accessed on April 18, 2024.

https://www.sbp.org.pk/reports/quarterly/fy19/Third/Special-Section-2.pdf. Accessed on April 11, 2024.

Chapter 13

Determinants and Regional Imbalance of Food Security for the Selected States in India

Saddam Hossain Mondal[a] *and Arup Kumar Chattopadhyay*[b]

[a]Department of Economics, Heramba Chandra College, India
[b]Department of Economics, The University of Burdwan, India

Abstract

Food security is the major issue of the world as well as in developing countries like India in recent times. The study examines the status of food security for 15 selected states in India during 1995–1996 to 2018–2019 where the statuses of food security for the states are estimated by applying the multidimensional food security index. The secondary data on different food security indicators are collected from the Ministry of Agriculture, Food and Agricultural Organization, National Family Health Survey (NFHS) report, RBI bulletin, Economic and Political Weekly Research Foundation (EPWRF), and different issues of Statistical Abstract of India. The value of the food security index has significantly increased over the study period. The food security conditions for all the states have improved at higher rates in the second subperiod compared to the first subperiod in the country. It has also used panel autoregressive distributed lag (ARDL) method for the estimation of log-term and short-term relationship between food security and socio-economic indicators of various states in the country. The food security of most of the states could be improved with the improvement in the conditions on the per capita power consumption, workforce participation rate, and cropping intensity in the country. Furthermore, the regional imbalance has decreased establishing that food security of the poorer states improved at a higher rate compared to richer states.

Keywords: Food insecurity; hunger; multidimensional food security index; panel autoregressive distributed lag model; regional imbalance

Growth, Poverty and Developmental Aspects of Agriculture, 191–210

Published under exclusive licence by Emerald Publishing Limited
doi:10.1108/978-1-83608-076-320241013

Introduction

An associated problem of sustaining steady growth in agriculture in a densely populated nation like India is achieving food security. The basic needs are the most important parameter for determining a person's ability to survive in the society. The basic needs of human life are food, cloth, water, shelter, and sanitation (Sok et al., 2018). Among these needs, food is the most essential need for human survival. The growth of the human body depends on various nutritious foods that provide necessary caloric intake and different types of vitamins, protein, and carbohydrates for a healthy life (Geissler & Powers, 2017). However, in recent times, food availability is the major issue worldwide. Worldwide, nearly 821 million people do not get sufficient food each, resulting in hunger (FAO, IFAD, UNICEF, WFP, & WHO, 2018). Apart from malnutrition, these people suffer from different types of diseases, such as anemia, heart disease, and poor immunity-related diseases (Resilience, 2017). Different world organizations such as Food and Agricultural Organization (FAO), World Bank (IBRD), Asian Development Bank, and International Monetary Fund (IMF) have taken various policies and conducted different meetings to overcome this food insecurity problem. These organizations/institutions provided various definitions of food security. In 1983, FAO gave the definition of food security as "ensuring that all people at all times have both physical and economic access to basic foods they need." World Development Report (1986) stated that food security is "access by all people at all times to enough food for an active, healthy life." FAO identified four components of food security, namely availability, accessibility, utilization, and stability to determine whether a population is food secured or not (Gross et al., 2000).

Availability: The availability component is the main pillar of food security. Availability represents the supply side of the food (McKay et al., 2019).

Accessibility: FAO and many economists, like A. K Sen, have emphasized that people are considered to be food secured if they have both physical and economic access to the food, which represents accessibility dimension of food security.

Utilization: The utilization dimension states that the food security of the people not only depends on the availability or accessibility of the food but also depends on the nutritional security of the people to be achieved from food utilization.

Stability: The stability dimension of food security represents that the people must consume safe and nutritious food consistently over the time or whole time.

Globally, most of the countries are facing a chronic food insecurity problem. Millennium development goals (MDGs) actively make target of the removal of food insecurity and hunger problem by 2015 from the world. In 1990, the food insecure population was 1.9 billion when the MDGs aimed at reducing this number to 836 million by 2015 (FAO, 2015). However, the MDGs were not fulfilled in achieving food security and also hunger to zero level in India. The United Nations again took a Sustainable Development Goal 2 (SDG2) for removing food insecurity and reducing the number of hungry people at zero level by 2030 (Gil et al., 2019). The shortage of food can be traced back to the traditional society, and it is still continuing during the 21st century. In the recent past, 26.1% of world population cannot consume the required minimum level of food in their dietary pattern. The condition of Africa is getting worst day by day. Here, more than half of the people face a chronic food insecurity problem. The food insecurity problem of Latin America has also increased in the recent past (FAO).

The food crisis problem of India is not that exists today; rather, it dates back to the British era in the nation. The British government's promotion of commercial farming and extraction of farm income from the farmers led to severe food insecurity problem in the nation. The nation faced an extremely severe food crisis after the Second World War (Pillay & Kumar, 2018). Since Independence, the country has been taking various steps for the improvement of food security; however, the issue has not been sorted out due to population pressure. India is currently one of the most food insecure countries in Asia. Nearly 22% of citizens cannot afford two full meals in a day due to lack of income in the nation (Asian Development Bank). India is producing a huge amount of food grains. In 2018, the country produced nearly 270 million tonnes of food grains . The net per capita food availability of the country is 484 grams in 2018, whereas the actual consumption of poorer is much less than this average. According to Tendulkar report from 2012, 25.7% of rural households and 13.7% of urban households lived below the poverty line in the country (Planning Commission, 2013). Poverty is still the major issue of the country; in addition, the food prices are increasing continuously, so the civilians below the poverty line are not being able to purchase essential food for their healthy life (Gustafson, 2013). In the recent past, however, the prices of the agricultural products that farmers received remain constant, while the costs of agricultural inputs are increasing rapidly, which has led to decline in the profitability of farmers (Narayanamoorthy, 2013). In particular, the low income of the marginal and small farmers is also one of the reasons for food insecurity in the nation. The rank of the Global Hunger Index of the country is steadily declining. The score of the Global Hunger Index decreased from 38.5 in 2005 to 30.3 in 2019 (Hilfe & Worldwide, 2019). The score of the index indicates that the country had a modest issue with food insecurity. If a physically fit Indian's intake is 2,250 kilocalories in an urban or rural setting, then he or she is not termed as a poor. India's per capita caloric intake capacity in 2012 was 2,233 kilocalories in rural areas and 2,206 kilocalories in urban areas, which indicates that both rural and urban areas in the country faced a deficit in food intake capacity, which was also evident from the protean fat consumption. A single rural person of the country consumed 56.5 grams of protein and 41.6 grams of fat, whereas the urban people on an average consumed 55.7 grams of protein per day and 52.5 grams of fat in 2012. (The data are collected from NSSO 68th round report.)

Malnutrition is a serious issue in India. Nearly 15.2% of people are suffering from undernourishment. These people are being affected by various health problems due to lack of nutrition. Anemia is the commonly observed disease in both adult males and females of the nation. From the National Family Health survey-4 (NFHS) reports, we see that nearly 53% of adult females are suffering from anemia disease. Indians are also facing a deficit in micronutrients, which results in various diseases such as rickets, scurvy, night blindness, etc. (IIPS, 2017).

Notable scholars like, Ahungwa et al. (2013), Butaumocho and Chitiyo (2017), Wineman (2014), Tagade (2011), Zhou et al. (2019), Sultana and Kiani (2011), Sisha (2020), Kumar et al. (2012), Jha et al. (2010), Singh and Bhogal (2008), and Owino et al. (2014), estimated the food security status at the household level and aggregated level of various states in the country and also outside of the country.

From these earlier research studies, we observe that the food security condition of various states in the country and in foreign countries increased in the recent past. After the 1980s, the country effectively implemented three-tier Panchayati Raj system all over the country. In 1991, the government also introduced Liberalization, Privatization, and Globalization policies in the country. It resulted to the effect of causing the nation's agricultural production and population's income level to rise at faster rates. Following the reform period, India has exported a significant quantity of food grains and horticultural crops to foreign countries and earned a huge amount of foreign currency . However, presently, we observe that more than 24% population has faced chronic food insecurity and malnutrition problem in the country due to lack of income and purchasing power. However, we have not come across any previous studies that measured the state-wise food security condition in the country by using a modified multidimensional food security index and its regional imbalance across the states in the country by using panel data regression. We also find that none of the previous studies examined short-term and long-term relationship between food security status and socioeconomic indicators of various states in the country. The study is a modest attempt to fill in these gaps in the literature.

Data and Methodology

Data

The study estimated the food security of several Indian states using secondary data at the state level. We took time series as well as panel data for the research study. The study period from 1995–1996 to 2018–2019 (i.e., the post-reform period) is taken into consideration for examining the state-wise food security conditions in India. The study period is divided into two subperiods: one is the early liberalization period (1995–1996 to 2006–2007), and the other is the matured liberalization period (2007–2008 to 2018–2019). Twelve indicators for creating a multidimensional food security index have been selected. Data on infant mortality rate, consumer price index of agricultural workers, per capita NSDP, cost of drinking water and sanitation, government cost of nutritional activities, cereal yield, per capita food production, and area of the food production, were collected from different government publications, namely Reserve Bank of India Bulletin, Statistical Abstract of India, different issues of census reports in India, NFHS reports, different reports of State Finance, and Economic and Political Weekly Research Foundation (EPWRF).

For estimating the relationship between food security and other socioeconomic indicators by using the panel autoregressive distributed lag (ARDL) model, we collected data on the agricultural wage rate of the woman workers, total length of the road, workforce participation rate, per capita electric consumption, cropping intensity, and government cost on education from various government publications and reports such as reports of Ministry of Labour and Employment, Bulletin of Reserve Bank of India, Publications of Directorate of Economics and Statistics, and different State Finance reports.

Methods

Normalization of Data

Data of the selected indicators of food security are given on a different scale. If we formulate the index based on different scales of dataset, we face the problem of getting the appropriate results of food security. To avoid this problem, we used the standardization method for the transformation of different scales of data to a standard scale before the application of principal component analysis (PCA).

Z Score is the common method for the standardization of the data whose mean value is zero and the standard deviation is one.

$$Z \text{ score of a particular indicator} = \frac{\text{Actual Value} - \text{Mean}}{\text{Standard Deviation}} \tag{1}$$

Furthermore, we also applied two most important diagnostic tests, namely Bartlett's test and Kaiser–Meyer–Olkin (KMO) test to check the adequacy and reliability or acceptability of the PCA.

The four dimensions of food security are availability, accessibility, affordability, and stability. We took several variables in each dimension. We applied PCA for reduction of the several variables into a single variable and then the formulation of an individual dimension index for food security of the various states in the country (estimated results of PCA reported in Table 13.1).

$$P_j = \frac{\sum_{i=1}^{I} Z_{ij} W_i}{\sum_{i=1}^{I} W_1} \tag{2}$$

where Z_{ij} represents the respective value of coefficient of j the states in the country, and W_{ij} represents the weight of jth variable taken from the corresponding element of the communality matrix; $i = 1,2,3,4$ and $j = 1,2,3$.

After obtaining the dimension index, we formulated the multidimensional food security index by using the equal weight method. The weight has been assigned equally to the four pillars of food security.

The general form of food security index is:

$$\begin{aligned}\text{Food Security Index} = {} & 0.25\,(\text{Food Availability}) + 0.25\,(\text{Food Accessibility}) \\ & + 0.25\,(\text{Food Utilization}) + 0.25\,(\text{Food Stability})\end{aligned}$$

We also used the UNDP goal post method for the transformation of index value into a comparable scale so that its value lies within the range 0–100 in the

Table 13.1. Estimated Results of Food Security in India.

States	Mean/CV	First Subperiod					Second Subperiod					Whole Period				
		AV	ACC	AFF	STA	MDFSI	AV	ACC	AFF	STA	MDFSI	AV	ACC	AFF	STA	MDFSI
Andhra Pradesh	Mean	47.38	22.46	23.49	51.16	36.36	65.61	55.13	66.08	48.03	62.68	56.50	38.80	44.79	49.59	49.52
	CV	36.70	26.30	26.99	21.19	11.87	24.65	37.93	30.31	23.48	19.78	33.41	57.88	58.41	22.05	32.76
Assam	Mean	48.40	30.66	21.63	50.09	38.30	64.59	46.93	67.95	49.10	60.74	56.50	38.80	44.79	49.59	49.52
	CV	28.39	29.99	42.40	20.99	10.52	23.34	42.78	24.96	18.40	20.89	28.94	44.81	60.63	19.36	29.69
Bihar	Mean	56.66	29.05	21.52	52.16	40.99	56.33	48.54	68.51	47.02	58.06	56.50	38.80	45.01	49.59	49.52
	CV	34.56	35.60	57.83	20.88	15.60	31.02	34.36	19.63	16.96	19.19	32.13	43.39	60.29	19.56	25.13
Gujarat	Mean	49.32	30.44	21.57	52.48	39.63	63.67	47.15	68.00	46.70	59.41	56.50	38.80	44.79	49.59	49.52
	CV	40.98	22.29	34.62	21.43	15.03	24.82	41.83	27.34	16.87	16.25	33.98	43.20	61.32	20.05	25.82
Haryana	Mean	52.94	18.94	20.79	21.97	21.86	60.05	58.65	68.79	77.22	77.21	56.50	38.80	44.79	49.60	49.54
	CV	27.32	26.06	52.89	47.40	47.60	23.12	30.49	16.69	20.37	20.41	25.37	61.87	59.99	19.50	62.87
Karnataka	Mean	55.90	32.05	22.07	51.09	41.46	57.10	45.54	67.51	48.09	57.58	56.50	38.80	44.79	49.59	49.52
	CV	43.31	24.89	35.87	24.07	19.69	31.88	48.41	27.07	20.07	14.77	37.09	45.41	60.26	62.68	23.40
Kerala	Mean	64.70	20.97	37.15	56.15	46.97	48.29	56.62	52.43	43.03	52.07	56.50	38.80	44.79	49.59	49.52
	CV	23.36	28.93	23.38	29.94	17.55	9.09	32.39	16.73	4.63	14.91	24.31	58.21	25.82	22.02	16.67
Madhya Pradesh	Mean	50.47	23.11	21.25	53.12	37.76	62.52	54.48	68.33	46.06	61.29	56.50	38.80	44.79	49.59	49.52
	CV	32.45	14.94	59.06	21.58	11.07	35.43	39.63	20.18	26.08	27.36	35.44	56.79	60.92	27.20	34.22
Maharashtra	Mean	60.76	23.39	21.32	51.46	39.93	52.23	54.20	68.26	47.72	59.11	56.50	38.80	44.79	49.59	49.52
	CV	28.97	33.78	58.82	22.28	12.75	37.69	38.39	19.34	18.30	16.95	33.23	56.75	60.46	24.27	25.25

Orissa	Mean	49.55	29.30	23.06	52.54	39.77	63.45	48.29	66.52	46.64	59.28	56.50	38.80	44.79	49.59	49.52
	CV	42.15	27.01	51.23	25.55	18.26	22.20	39.04	18.48	17.55	15.99	33.30	44.20	56.11	20.46	26.13
Punjab	Mean	50.80	19.30	21.35	53.43	36.77	62.19	58.29	68.23	45.75	62.27	56.50	38.80	44.79	49.59	49.52
	CV	32.67	56.10	38.45	17.88	28.15	20.94	20.45	24.80	14.16	13.96	27.80	58.82	60.84	22.75	32.37
Rajasthan	Mean	49.91	19.44	21.46	56.29	37.99	63.08	58.15	68.12	42.90	61.06	56.50	38.80	44.79	49.59	49.52
	CV	47.79	29.39	53.59	28.30	20.73	17.95	31.60	22.39	16.42	16.53	34.44	61.44	60.83	17.94	29.76
Tamil Nadu	Mean	57.27	32.45	20.88	57.86	44.92	55.73	45.14	68.70	41.33	54.13	56.50	38.80	44.79	49.59	49.52
	CV	41.31	27.37	55.06	23.11	11.13	38.36	43.53	18.99	14.51	21.25	39.05	41.91	60.78	27.93	19.92
Uttar Pradesh	Mean	62.64	28.07	23.12	50.74	42.15	50.35	49.52	66.46	48.44	56.90	56.50	38.80	44.79	49.59	49.52
	CV	30.77	36.06	48.72	18.17	16.95	29.75	30.44	25.31	18.01	20.10	15.88	42.96	58.48	26.60	24.20
West Bengal	Mean	60.72	28.88	21.80	47.38	39.89	52.27	48.72	67.78	51.80	59.16	56.50	38.80	44.79	49.59	49.52
	CV	12.69	38.15	61.44	15.59	19.90	16.02	35.80	19.67	24.64	16.47	24.31	45.10	60.00	17.86	26.52
India	Mean	49.32	24.67	24.32	53.61	39.06	63.67	52.92	65.26	45.57	59.99	56.50	38.80	44.79	49.59	49.52
	CV	25.95	29.71	53.01	21.07	12.11	17.08	18.66	21.69	12.56	15.53	31.88	43.17	55.26	21.06	26.06

Source: Authors' calculations based on secondary data.

Note: AV = Availability, ACC = Accessibility, AFF = Affordability, STA = Stability, MDFSI = Multidimensional Food Security Index.

country: 0 represents that the food security conditions are very poor, and 100 represents maximum (i.e., perfect) food security in the country.

$$\text{Food Security Index Score} = \frac{\text{Actual value–Minimum value}}{\text{Maximum value–Minimum value}} \times 100 \quad (3)$$

Econometric Model for Determinants of Food Security

Food insecurity is the major issue in recent times in the country as well as in the world. The food security of the country depends on numerous socioeconomic factors. We selected a few important socioeconomic indicators, as per availability, to check whether these negatively or positively impact the food security condition across the selected states in the country.

Panel Cointegration Tests

First, we used Fisher Chi-square Augmented Dickey–Fuller test (ADF) panel unit root test and determined the order of integration of dependent and independent variables. It is observed that all the variables are either I(0) or I(1), so that we estimated panel ARDL model. After confirming the order of integration, we conducted the Pedroni panel cointegration test (1999, 2004) for checking the existence of the long-term relationship between dependent variable, i.e., score of food security index, and the selected independent variables, such as work force participation rate, government cost on education, cropping intensity, and agricultural wage rate of woman workers, and so on, across the states in the country. The null hypothesis of the Pedroni cointegration test indicates that there is no cointegration among the variables concerned, and the alternative hypothesis represents that there is a presence of cointegration. For Pedroni cointegration test, the following equation is estimated (notation having their usual meaning):

$$y_{it} = x_{it}^{'}\beta_i + z_{it}^{'}\tau_i + \epsilon_{it} \quad (4)$$

Panel ARDL

In the study, we estimated the panel ARDL model after checking the order of integration of dependent and independent variables and also the existence of a long-term relationship between them. The panel ARDL model separately clarifies the presence of short-term and long-term relationship between the dependent and independent variables.

The panel ARDL equation is represented as follows:

$$\begin{aligned} Y_{it} = {} & \alpha_i + \sum_{j=1}^{p}\alpha_{1,ij}Y_{i,t-j} + \sum_{j=0}^{q_1}\alpha_{2,ij}D_{i,t-j} + \sum_{j=0}^{q_2}\alpha_{3,ij}L_{i,t-j} \\ & + \sum_{j=0}^{q_3}\alpha_{4,ij}H_{i,t-j} + \sum_{j=0}^{q_4}\alpha_{5,ij}K_{i,t-j} + \sum_{j=0}^{q_5}\alpha_{6,ij}M_{i,t-j} + \epsilon_{it} \end{aligned} \quad (5)$$

where p indicates the lag length of the dependent variable, and q_s represents the lag length of the sth independent variables ($s = 1,2,3 \ldots 5$). $i = 1, 2, 3, \ldots N$ and $t = 1, 2, 3, \ldots T$, α_i is the fixed effects, α_1 to α_5 represent the coefficients of the lagged independent variables, and ϵ_{it} is the error term. The independent variables are defined later.

For short-term relationship, the panel error correction (ECM) equation corresponding to ARDL is given by:

$$\begin{aligned} \Delta Y_{it} = \alpha_i &+ \sum_{j=1}^{p}\alpha_{1,ij}\Delta Y_{i,t-j} + \sum_{j=0}^{q_1}\alpha_{2,ij}\Delta D_{i,t-j} + \sum_{j=0}^{q_2}\alpha_{3,ij}\Delta L_{i,t-j} \\ &+ \sum_{j=0}^{q_3}\alpha_{4,ij}\Delta H_{i,t-j} + \sum_{j=0}^{q_4}\alpha_{5,ij}\Delta K_{i,t-j} + \sum_{j=0}^{q_5}\alpha_{6,ij}\Delta M_{i,t-j} \\ &+ \sum_{j=0}^{q_6}\alpha_{7,ij}\Delta N_{i,t-j} + \beta_{1,ij}Y_{i,t-1} + \beta_{2,ij}D_{i,t-1} + \beta_{3,ij}L_{i,t-1} \\ &+ \beta_{4,ij}H_{i,t-1} + \beta_{5,ij}K_{i,t-1} + \beta_{6,ij}M_{i,t-1} + \beta_{7,ij}N_{i,t-1} + \epsilon_{it} \end{aligned} \tag{6}$$

where Δ represents the first difference of variable concerned. Here, α_1 to α_7 indicate the short-term coefficients, and β_1 to β_7 represent the long-term coefficients of the score of food security (Y), agricultural wage rate of woman workers (D), per capita consumption of electricity (L), cropping intensity (H), work force participation rate (K), government cost on education (M), and total length of the road (N), respectively.

The long-term relationship is established between the dependent variable and independent variables with the panel ECM model (eq. 6) as follows:

$$\begin{aligned} \Delta Y_{it} - \alpha_i &+ \sum_{j=1}^{p}\alpha_{1,ij}\Delta Y_{i,t-j} + \sum_{j=0}^{q_1}\alpha_{2,ij}\Delta D_{i,t-j} + \sum_{j=0}^{q_2}\alpha_{3,ij}\Delta L_{i,t-j} \\ &+ \sum_{j=0}^{q_3}\alpha_{4,ij}\Delta H_{i,t-j} + \sum_{j=0}^{q_4}\alpha_{5,ij}\Delta K_{i,t-j} + \sum_{j=0}^{q_5}\alpha_{6,ij}\Delta M_{i,t-j} \\ &+ \sum_{j=0}^{q_6}\alpha_{7,ij}\Delta N_{i,t-j} + \theta_i \mathrm{ECM}_{i,t-1} + \epsilon_{it} \end{aligned} \tag{7}$$

where θ_i represents the coefficient of the ECM measuring the speed of adjustment that is made every year toward long-term equilibrium.

The optimal lag length of the ARDL models (eqs. 5–7) is determined based on Akaike's lag selection criterion, and a maximum lag length is restricted to four keeping in mind the limited number of annual observations.

There are mainly three types of estimations for panel ARDL regression process, viz. pooled mean group (PMG), mean group (MG), and difference fixed effect (DFE) model. We have used Housman model specification test for selecting the appropriate method. This test argued that the PMG is the appropriate method for the estimation of panel ARDL regression. This panel ARDL regression was used for the estimation of both long-term and short-term relationship by Pesaran et al. (1997, 1999). They combined both averaging and pooling of coefficients.

Measurement of Convergence in Food Security

We used the coefficient of variation (CV) method for the estimation of the cross-sectional dispersion of the food security for the selected states over the study period in the country.

The absolute beta convergence shows that the growth rates of logarithmic variables are negatively related to the initial value of the variables, i.e., poor countries grow at a faster rate compared to rich countries. We used Fisher Chi-Square Phillips Perron test method for the estimation of absolute beta convergence across the states in the country.

Results and Discussion

Status of Food Security in India

The conditions of food security have been divided into four categories on the basis of index scores as follows:

(1) $0 \leq \text{MDFSI} < 25$: Poor Food Security.
(2) $25 \leq \text{MDFSI} < 50$: Satisfactory Food Security.
(3) $50 \leq \text{MDFSI} < 75$: Good Food Security.
(4) $\text{MDFSI} \geq 75$: Excellent Food Security.

We divided the variability of food security into three ways on the basis of CV values:

(1) Low variability whose CV lies between 0 and 15.
(2) Medium variability whose CV lies between 15 and 30.
(3) Wide variability whose CV is above 30.

From Table 13.1, we observe that the food security condition lies in the category of satisfactory level with medium variability in Assam followed by Bihar, Gujarat, Karnataka, Kerala, Maharashtra, Orissa, Rajasthan, Tamil Nadu, Uttar Pradesh, West Bengal, and India as a whole over the whole study period, while a few other states, namely Andhra Pradesh, Haryana, Madhya Pradesh, and Punjab achieved a satisfactory level of food security with wide variability. The food security condition is observed at a satisfactory level across all the selected states as well as in the whole of India because of the poor performance of accessibility dimension over the study period. It not only reduced the food security condition but also increased the variability in food security in all the cases. The affordability dimension is also one of the response variables for the high variability in food security across the states.

In the case of the first subperiod, we note (from Table 13.1) that the food security condition of the country and of the selected states except for Haryana is found to be at satisfactory level. Among those states, food security belongs to the category of satisfactory level with low variability in Andhra Pradesh, Assam, Madhya Pradesh, Maharashtra, Tamil Nadu, and India, whereas it belongs to the category of satisfactory level with medium variability in nine other states, i.e., Bihar, Gujarat, Karnataka, Kerala, Orissa, Punjab, Rajasthan, Uttar Pradesh, and West Bengal in the first subperiod. Haryana is the only state whose food security condition is found to be poor with high variability during the first subperiod. In the first subperiod, the food security condition is significantly influenced by the poor performance of accessibility and affordability dimensions among the states and the whole India. However, in the case of the variability in food security, we observe that it increased mainly due to variation in both availability and affordability dimensions.

In the case of the second subperiod (i.e., during matured liberalization period), we observe that the food security condition improves in all the selected states as well as in the whole country compared to that either in the first subperiod or over the whole study period. The food security condition exists in a good category with low variability in Karnataka, Kerala, and Punjab in the second subperiod. The rest of the other states and the country as a whole are found to have good food security situation with medium variability in the second subperiod. Haryana turns its picture and plays an excellent role with medium variability in food security condition during this period. The fluctuations in food security are mainly observed due to the variation in accessibility dimension in the second subperiod.

Estimated Results of Panel Unit Root Test

It is noted that (Table 13.2) net irrigated area is the only explanatory variable that is found to be stationary, i.e., I(0) at level. However, in the cases of other regressors, i.e., index of food security, agricultural wage rate of woman workers, government cost on education, workforce participation rate, per capita consumption of electricity, and total length of the road, we observe that these are nonstationary at level. Therefore, when we take the first difference of these selected series, those are found to be statistically significant at 1% level, i.e., those are I(1). After insertion of trend in the equation for unit root test, we observe that the four variables, i.e., agricultural wage rate of woman workers, government cost on education, workforce participation rate, and total length of the road are each nonstationary at levels but stationary at first differences, i.e., I(1). On the other hand, in the cases of food security index, net irrigated area, and per capita consumption of electricity, we see that these three series are stationary [i.e., I(0)] at level. Thus, from the table, we observe that the dependent variable and regressors are stationary either at level of first difference, i.e., either I(0) or I(1).Therefore, the panel ARDL model is here appropriate for examining the short-term and long-term relationship between dependent variable and regressors over time across the states in the country.

Estimated Results of Panel Cointegration Test

Panel cointegration test establishes the existence of long-term relationship between dependent and independent variables. We used Pedroni cointegration test for checking the existence of long-term relationship between dependent variable and regressors. The null hypotheses of the tests represent that there is no cointegration, i.e., there is no long-term relationship between dependent and independent variables. From Table 13.3, we observe that both cointegration tests are statistically significant at 1% level, i.e., the null hypothesis of no cointegration in the panel is rejected; hence, there exists a long-term relationship in the panel.

Table 13.2. Estimated Results of Panel ADF Fisher Chi-Square Unit Root Test During the Period 1995–1996 to 2018–2019.

	Level		Trend	
Variable Name	**I(0)**	**First Difference: I(1)**	**I(0)**	**First Difference: I(1)**
Index value of food security	25.570 (0.696)	662.212*** (0.000)	119.322*** (0.000)	NA
Agricultural wage rate of women	0.079 (1.000)	87.751*** (0.000)	2.075 (1.000)	96.435*** (0.000)
Cost of education	0.0108 (1.000)	73.072*** (0.000)	0.824 (1.000)	179.096*** (0.000)
Per capita consumption of electricity	1.825 (1.000)	260.647*** (0.000)	45.358** (0.035)	NA
Total length of the road	26.714 (0.638)	216.360*** (0.000)	29.255 (0.5042)	163.257*** (0.000)
Net irrigated area	229.557*** (0.000)	NA	219.835*** (0.000)	NA
Work force participation rate	17.443 (0.966)	284.984*** (0.000)	35.075 (0.239)	206.002*** (0.000)

Source: Authors' calculations based on secondary data.

Note(s): *** Significant at 1% level, ** Significant at 5% level, * Significant at 10% level.

Estimated Results of PMG Regression Based on Full Panel

We used PMG regression method and estimated the long-term and short-term relationship between food security and its selected explanatory variables over the study period in the country. First, we estimated the long-term and short-term relationship of full panel data, and later, we analyzed the state-wise short-term relationship between the food security and its other responsible variables. The lag length of the estimation in each case is determined by the Akaike information criterion (AIC) lag selection criterion.

Long-term Analysis

From Table 13.3, we observe that the food security and its explanatory variables are significantly related to each other across the full panel in the long term. From Table 13.3 specifically, we observe that the government cost on education, agricultural wage rate of woman workers, total length of the road, net irrigated area, and workforce participation rate are statistically significant with positive value, i.e., the positive association between food security and the regressors exists in the country in long term. More specifically, a 1% partial increase in each of agricultural wage rate of woman workers, total length of the road, net irrigated area, government cost on education, and workforce participation rate will have improvement of food security, respectively, by 0.004%, 0.070%, 0.003%, 0.075E-8%, and 0.0070% in the long term. We also observe that the per capita consumption of electricity is statistically insignificant in affecting food security in the country, i.e., there is no association between food security and per capita power consumption in the long term.

Short-term Analysis

Table 13.3 represents that the error correction value is statistically significant and negative (−0.825). The statistically significant with negative ECM indicates that there exists a long-term relationship between food security and explanatory variables in the selected states of the country, and the short-term disequilibrium situation, if any, would be corrected by 82.5% in each year in the long term. We also observe that the changes in government cost on education and workforce participation rate are positively related to the food security in the short term. The improvement by 1% in change in government cost on education and workforce participation rate would have an improvement in food security in the country in short-term by 0.57E-7% and 0.008%, respectively. The rest of other explanatory variables, namely agricultural wage rate of woman workers, total length of the road, net irrigated area, and per capita consumption of electricity, are found to be statistically insignificant in the short-term.

Estimated Results of State-specific Panel Regression in Short Term

One advantage of PMG estimation of panel ARDL model is to know the unit specific (here state specific) regression results in the short term. This section is devoted to analyze these disaggregated short-term results. From the state-specific panel ARDL results, we indicated here only relevant factors that had an impact on the food security condition in the country during the study period.

From Table 13.4, we observe that the error correction coefficient is statistically significant with a negative value in all the states except Haryana and Kerala. It implies that the short-term disequilibrium would be corrected to establish the long-term equilibrium with the improvement in selected explanatory variables in most of the states in the country. On the contrary, in the cases of Haryana and Kerala, we observe that the short-term disequilibrium situation prevails even in the long term because the estimated coefficients of ECM here are found to be statistically insignificant.

Government cost on education is an important factor for the improvement of social welfare. If the government cost on education (i.e., expenditure on training, skill development programs, improvement of educational institutes, faculty development programs, etc.) increases, then the capability of the people to earn

Table 13.3. Estimated Results of Panel Autoregressive Distributed Lag (ARDL) Model and Pedroni Cointegration Test.

	Variable Name	Coefficient	*t*- Statistic	Probability
Long-term	Cost of education	0.000000075**	2.32	0.020
	Agricultural wage of women worker	0.004128***	5.27	0.000
	Per capita electric consumption	−0.000243	−1.33	0.183
	Total length of the road	0.07037**	2.83	0.047
	Net irrigated area	0.003426*	1.61	0.087
	Workforce participation rate	0.007266***	3.47	0.001
Short-term	Error correction coefficient	−0.82589***	−7.30	0.000
	Δ Cost of education	0.00000057***	3.38	0.001
	Δ Agricultural wage of women worker	−0.002170	−0.97	0.332
	Δ Per capita electric consumption	0.00075	1.16	0.245
	Δ Total length of the road	−0.04560	−0.38	0.705
	Δ Net irrigated area	−0.000224	−0.85	0.394
	Δ Workforce participation rate	0.00858**	1.71	0.047
	Constant	−0.18935***	−4.04	0.000
Pedroni panel cointegration test		Statistic	Probability	
	Modified Phillips–Perron test	3.0453***	0.0012	
	Phillips–Perron test	−10.5702***	0.0000	
	Augmented Dickey–Fuller test	−9.5492***	0.000	

Source: Authors' calculations based on secondary data.

Note(s): *** Significant at 1% level, ** Significant at 5% level, * Significant at 10% level.

Table 13.4. Summary Results of State Specific Panel.

States	ECT	CE	AWWW	PCEC	TLR	NIA	WFPR
Andhra Pradesh	Negative significant	Positive insignificant	Positive significant	Positive insignificant	Positive significant	Negative insignificant	Negative significant
Assam	Negative significant	Positive significant	Positive significant	Positive insignificant	Negative insignificant	Negative insignificant	Negative insignificant
Bihar	Negative significant	Positive significant	Positive insignificant	Positive insignificant	Positive insignificant	Positive significant	Positive insignificant
Gujarat	Negative significant	Positive insignificant	Negative insignificant	Positive significant	Negative insignificant	Negative significant	Positive significant
Haryana	Negative insignificant	Positive insignificant	Negative insignificant	Positive insignificant	Negative significant	Positive insignificant	Positive insignificant
Karnataka	Negative significant	Positive significant	Negative insignificant	Negative insignificant	Positive insignificant	Positive insignificant	Positive insignificant
Kerala	Positive insignificant	Negative insignificant	Positive significant	Negative significant	Negative insignificant	Negative insignificant	Negative insignificant
Madhya Pradesh	Negative significant	Positive significant	Negative insignificant	Positive significant	Negative insignificant	Negative significant	Positive insignificant
Maharashtra	Negative significant	Positive insignificant	Negative insignificant	Positive significant	Negative insignificant	Negative insignificant	Positive significant
Orissa	Negative significant	Positive significant	Positive insignificant	Negative insignificant	Positive insignificant	Positive significant	Negative insignificant
Punjab	Negative significant	Positive insignificant	Positive insignificant	Positive insignificant	Negative insignificant	Positive insignificant	Positive significant

(Continued)

Table 13.4. *(Continued)*

States	ECT	CE	AWWW	PCEC	TLR	NIA	WFPR
Rajasthan	Negative significant	Positive insignificant	Negative insignificant	Positive insignificant	Negative insignificant	Negative insignificant	Positive significant
Tamil Nadu	Negative significant	Positive insignificant	Negative significant	Negative significant	Negative significant	Positive significant	Negative insignificant
Uttar Pradesh	Negative significant	Positive insignificant	Negative significant	Positive significant	Negative significant	Positive insignificant	Positive significant
West Bengal	Negative significant	Positive insignificant	Negative significant	Positive significant	Negative significant	Positive insignificant	Positive insignificant

Source: Authors' calculations based on secondary data.

Note: CE = Cost of Education, AWWW = Agricultural Wage of Women Worker, PCEC = Per Capita Electric Consumption, TLR = Total Length of the Road, NIA = Net Irrigated Area, WFPR = Workforce Participation Rate.

more income also increases, which leads to improvement in the food security condition of people. We observe that the government cost on education is significantly positively related to the food security in Assam, Bihar, Karnataka, Madhya Pradesh, and Orissa. However, we observe that the agricultural wage rate of woman workers has positively affected the food security condition in Andhra Pradesh, Assam, and Kerala in the country. The expansion of total length of road is also statistically significant with positive value in only one state, i.e., Andhra Pradesh, while for few other states, namely Haryana, Tamil Nadu, Uttar Pradesh, and West Bengal, it plays a statistically significant role with negative value in the country. In the case of per capita consumption of electricity, we observe that there exists a significantly positive relationship between food security and per capita consumption of electricity in Gujarat, Madhya Pradesh, Maharashtra, Uttar Pradesh, and West Bengal in the short term. The workforce participation rate also significantly positively affects the food security conditions of Gujarat, Maharashtra, Punjab, Rajasthan, and Uttar Pradesh in the short term. Andhra Pradesh is the only one of the states where it is astonishingly found to be statistically significant with a negative value and that happens positively because of wide sampling fluctuation in dataset. In the case of net irrigated area, we observe that there is a presence of significant direct short-term relationship between food security and net irrigated areas in Bihar, Orissa, Tamil Nadu, and Karnataka in the country. For rest of the states, this relation remains insignificant in the country during short term.

The Regional Disparity in Food Security in India

The CV with respect to food security has continuously declined across the states over the study period (Fig. 13.1), thus representing the presence of sigma convergence. The value of CV has actually decreased from 40.10 in 1995–1996 to 12.78 in 2018–2019. The estimated values of CV represent a clear picture of the decreasing trend in variability in food security over the study period.

We used Fisher Chi-Square Phillips–Perron (PP) test for measuring the absolute beta convergence of food security across the states in the country. From the estimated results, we observe that the food security has converged to its steady-state level across the states in the country. The value of Fisher Chi-Square Phillips Perron (PP) test statistic is 200.71, which is statistically significant at a 1% level. More specifically, the food security condition improves at a faster rate in the poorer states compared to the developed states during the study period in the country, which signifies the presence of also absolute beta convergence.

Conclusion

Over the study period and its subperiods, food security in India has been notably achieved with low variability across the states as well as in the country as a whole. The food security condition considerably improved in the second subperiod in all the states compared to the first subperiod, but a large number of states performed

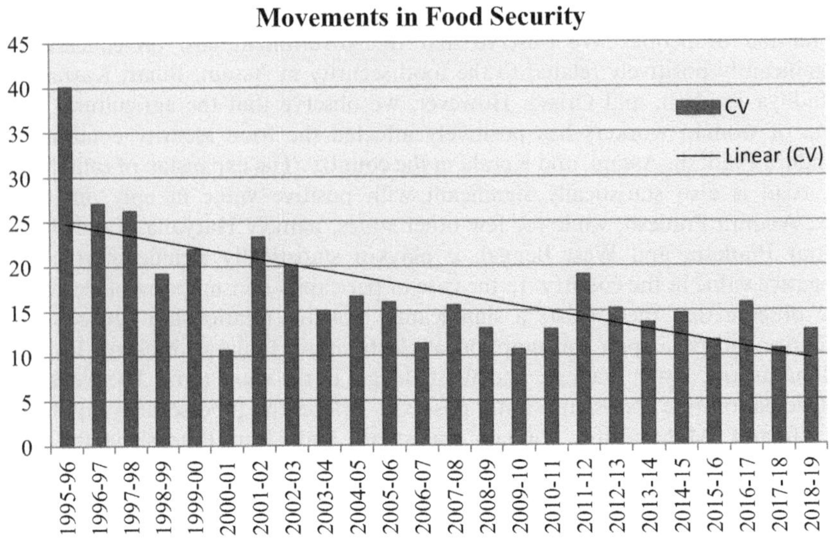

Fig. 13.1. Inter-state Disparity in Food Security in 15 States During 1995–1996 to 2018–2019. *Source:* Authors' calculations based on secondary data.

at a slower space over time due to poor performances in its accessibility and affordability dimensions in the second subperiod. The food security and the qualitative food consumption pattern of the households in the country are not impressive in the recent past. Therefore, it can be concluded that the food security situation considerably improved but is associated with the nutritional deficiency in the country. We also observe that the food security of different states deteriorated mainly due to poor economic accessibility. If the government works to increase accessibility and the purchasing power of the people , then the food security will significantly improve in the future. The workforce participation rate and government cost on education are directly related to food security in most of the cases in the country. If the government takes different job-oriented policies, promotes new industries, and encourages new training and skill development programs, the income level of the population and hence their food security will significantly improve in the country. The regional imbalance decreased, demonstrating that food security of the poorer states improved more quickly than that of the richer states (i.e., existence of catch-up effect) in the country.

References

Ahungwa, G. T., Umeh, J. C., & Muktar, B. G. (2013, November–December). Empirical analysis of food security states of farming households in Benue state, Nigeria. *IOSR Journal of Agriculture and Veterinary Science*, *6*(1), 57–62.

Butaumocho, B., & Chitiyo, P. T. (2017). A comparative analysis of household food security measures in rural Zimbabwe. *The International Journal of Food and Agricultural Economics*, *5*(2), 41–58.

FAO, IFAD, UNICEF, WFP, & WHO. (2018). *The state of food security and nutrition in the world 2018. Building climate resilience for food security and nutrition.* FAO.

Food and Agriculture Organization of the United Nations. (2015). *The state of food insecurity in the world 2014: Strengthening the enabling environment for food security and nutrition.* Food & Agriculture Org.

Geissler, C., & Powers, H. J. (Eds.). (2017). *Human nutrition.* Oxford University Press.

Gil, J. D. B., Reidsma, P., Giller, K., Todman, L., Whitmore, A., & van Ittersum, M. (2019). Sustainable development goal 2: Improved targets and indicators for agriculture and food security. *Ambio*, *48*(7), 685–698.

Gross, R., Schoeneberger, H., Pfeifer, H., & Preuss, H. J. (2000). The four dimensions of food and nutrition security: Definitions and concepts. *SCN News*, *20*(20), 20–25.

Gustafson, D. J. (2013). Rising food costs & global food security: Key issues & relevance for India. *Indian Journal of Medical Research*, *138*(3), 398–410.

Hilfe, W. H., & Worldwide, C. (2019). *Global Hunger Index. The challenge of hunger and climate change.* https://www.globalhungerindex.org/pdf/en/2019.pdf

IIPS, O. (2017). *National family health survey (NFHS-4): 2014-15. India.* International Institute of Population Sciences.

Jha, R., Bhattacharya, S., & Gaiha, R. (2010). Social safety nets and nutrient deprivation: An analysis of the national rural employment guarantee program and the public distribution system in India. *Journal of Asian Economics*, *22*.

Kumar, A., Bantilan, M. C. S., Kumar, P., Kumar, S., & Jee, S. (2012). Food security in India: Trends, patterns and determinants. *Indian Journal of Agriculture Economics*, *67*(3).

McKay, F. H., Haines, B. C., & Dunn, M. (2019). Measuring and understanding food insecurity in Australia: A systematic review. *International Journal of Environmental Research and Public Health*, *16*(3), 476.

Narayanamoorthy, A. (2013). Profitability in crops cultivation in India: Some evidence from cost of cultivation survey data. *Indian Journal of Agricultural Economics*, *68*(1), 104–121.

Owino, A., Wesonga, R., & Nabugoomu, F. (2014). Determining food insecurity: An application of the Rasch model with household survey data in Uganda. *International Journal of Food Science*, *3*. https://doi.org/10.1155/2014/121269

Pedroni, P. (1999). Critical values for cointegration tests in heterogeneous panels with multiple regressors. *Oxford Bulletin of Economics & Statistics*, *61*(S1), 653–670.

Pedroni, P. (2004). Panel cointegration: Asymptotic and finite sample properties of pooled time series tests with an application to the PPP hypothesis. *Econometric Theory*, *20*(3), 597–625.

Pesaran, M. H., Shin, Y., & Smith, R. P. (1997). *Pooled estimation of long-run relationships in dynamic heterogeneous panels.* DAE working papers amalgamated Series, 9721.

Pesaran, M. H., Shin, Y., & Smith, R. P. (1999). Pooled mean group estimation of dynamic heterogeneous panels. *Journal of the American Statistical Association*, *94*(446), 621–634.

Pillay, D. P. K., & Kumar, T. M. (2018). Food security in India: Evolution, efforts and problems. *Strategic Analysis*, *42*(6), 595–611.

Planning Commission. (2013). *Press note on poverty estimates, 2011–12* (No. id: 5421).

Resilience, B. U. I. L. D. I. N. G. (2017). The State of food security and nutrition in the world. In *Rome: Building resilience for peace and food security*.

Singh, G., & Bhogal, T. S. (2008). Food security: Key issues and strategies for Kerala. A note. *Indian Economic Review*, *43*(1), 141–147.

Sisha, T. A. (2020). Household level food insecurity assessment: Evidence from panel data, Ethiopia. *Scientific African*, *7*. https://doi.org/10.1016/j.sciaf.2019.e00262

Sok, P., Gardner, S., Bekele, T., Globerman, J., Seeman, M. V., Greene, S., Sobota, M., Koornstra, J. J., Monette, L. V., Hambly, K., Hwang, S. W., Watson, J., Walker, G., & Rourke, S. B. (2018). Unmet basic needs negatively affect health-related quality of life in people aging with HIV: Results from the positive spaces, healthy places study. *BMC Public Health*, *18*(1), 1–10.

Sultana, A., & Kiani, A. (2011). Determinants of food security at household level in Pakistan. *African Journal of Business Management*, *5*(34), 12972–12979.

Tagade, N. (2011). *Food security in Maharashtra: Regional dimension*. The Institute Of Social and Economic Change.

Wineman, A. (2014, July 27–29). Multidimensional household food security measurement in rural Zambia. In *Selected papers prepared for presentation at the agricultural and applied economics associations 2014*. AAEA Annual Meeting.

Zhou, D., Shah, T., Ali, S., Ahmad, W., Din, I. U., & Ilyas, A. (2019). Factors affecting household food security in rural northern hinterland of Pakistan. *Journal of the Saudi Society of Agricultural Sciences*, *18*(2), 201–210.

Chapter 14

Food Security in Pakistan and Afghanistan: Investigating the Lacunae of Domestic Policy and Foreign Policy

Debasish Nandy

Kazi Nazrul University, India

Abstract

Pakistan and Afghanistan have long been suffering from a food crisis. Due to the acute fiscal crisis since 2021, both countries are in search of a way out of this economic hardship. Issues such as lack of industrialization, ethnic conflicts, political violence, terrorism, corruption, misgovernance, arms trafficking, drug trafficking and human trafficking, poverty, illiteracy, and overdependency on foreign aid are common in both countries. Pakistan has an additional feature of military involvement in politics and the decision-making process. Due to wrong governmental policy, corruption, lack of agriculture, or food security policy, food security is a fundamental issue for both the countries. Foreign currency reserves in Pakistan are currently depleted, and international funding agencies are not interested in sanctioning additional loans. The economic profile of the country has declined due to terror financing. Millions of people in Pakistan are experiencing food crisis, with starvation becoming more common. The government has no adequate money to import food grains from overseas. Since August 2021, the foreign-aid-dependent Afghan economy has been facing severe crisis . After the Taliban took over the administration of Afghanistan for the second time, foreign donors stopped all types of assistance, thereby intensifying the food crisis. Due to the shortage of water, modern irrigation systems, droughts, and modern agriculture equipment, Afghanistan is unable to produce adequate food. This chapter intends to investigate the causes and consequences of food security in Pakistan and Afghanistan through the content analysis method.

Growth, Poverty and Developmental Aspects of Agriculture, 211–224

Published under exclusive licence by Emerald Publishing Limited
doi:10.1108/978-1-83608-076-320241014

Keywords: Food security; Pakistan; Afghanistan; corruption; misgovernance; foreign aid

Introduction

Both Pakistan and Afghanistan have faced accusations of failing to provide food security for their respective citizens. The economically challenged, productively failed, politically unstable, socially fragmented, and industrially backward countries like Pakistan and Afghanistan have witnessed hunger and malnutrition, which are related to human security. Food security has been one of the key components of human security. In the UNDP (1994) Report, a fresh concept of human security is presented, shifting the focus from borders and weapons to development and individuals as the foundation of security. The notion of human security is thoroughly explored, considering both national and international factors. In the case of Pakistan, famine and food crisis are due to a lack of good governance, democratic deficiency, the militarization of the state system, spending on the military head, religious atrocities, unemployment, militancy, industrial backwardness, reckless corruption, the financial crisis of the government, drought, and other natural calamities. The overdependency on foreign aid and misuse of aid are also equally responsible for the food insecurity in Pakistan. Afghanistan also faces almost similar challenges, except militarization of the system. However, Afghanistan faces an unprecedented systematic crisis due to the Taliban administration, which hardly cares about human security. The unaccountable Taliban government is not only a financial loser but also internationally unrecognized. Therefore, foreign-aid-based supply chain has been interrupted since August 15, 2021. A large number of people of Afghanistan are in starvation, which is an immense threat to human security.

Ensuring the provision of food for every individual in a population using sustainable methods of food production, distribution, and consumption, food security encompasses the right of individuals to determine their policies and techniques. Emphasizing the importance of self-reliance through domestic food production, the term "food security" denotes the availability of an adequate amount of food to fulfill the energy requirements of a diet (Pinstrup-Andersen, 2009). However, it is important to note that the mere availability of food does not automatically guarantee access to it. As highlighted by Ahmad and Farooq (2010), food security is a flexible concept as reflected by the many attempts to define it in research and policy usage. The concept of food security originated in the 1970s, at a time of global food crises. Even two decades ago, there were about 200 definitions for food security in published writings. Maxwell and Frankenberger (1992, pp. 1–274) showed the contextual dependent features of the definition. The last revision to this definition happened at the 2009 World Summit on Food Security, which added a fourth-dimension stability as the short-term time indicator of the ability of food systems to withstand shocks, whether natural or man-made.

Challenges related to food insecurity can arise at any point along the pathways of availability, accessibility, utilization, and stability of food security. Coping strategies can be implemented at the family, personal, or governmental level to address these challenges. The interplay between stress and potentially inadequate coping strategies creates recurrent feedback loops, necessitating modifications in coping strategies until food security is regained (Peng & Berry, 2019). While sufficient food may be available at the national level, certain regions face significant challenges in terms of food availability due to prolonged droughts, social unrest, or limited arable land. This is especially true for many regions in Pakistan. Moreover, these locations also face issues with food distribution. On the other hand, access is defined as the ability of households or individuals to produce, purchase, or acquire suitable and nourishing food (Timmer, 2000).

Research Objectives, Research Questions, and Research Methodology

The two objectives of the study are (1) to delineate the causes and roots of food insecurity in Pakistan and Afghanistan and (2) to investigate the domestic and international initiatives of mitigating food security. There are two research questions in this chapter: (1) What are the primary challenges of food security in Pakistan and Afghanistan? and (2) What domestic and international initiatives are taken toward food insecurity in Pakistan and Afghanistan? The content analysis method has been applied in this chapter. Both primary and secondary sources are used in this chapter. Various reports, books, journals, and newspaper sources have been used in this chapter.

Review of Literature

In this study, the author reviewed the existing literature, an overview of the literature is given here. Anees (2023) in her article described Pakistan's economic crisis. In a report, Asian Development Bank (ADB) (2022) described the social protection and food security challenges in Pakistan. Ahmad and Farooq (2010) described the food security in Pakistan. Ahmed (2011) critically analyzed economic crisis of the state. Faiez (2024) observed in the Washington Post about the foreign aid drops in Afghanistan. Peng and Berry (2019) critically discussed the concept of food security. Khan and Shah (2010) analyzed food insecurity in Pakistan. Khetran (2023, pp. 1–6) critically analyzed the food insecurity in Pakistan. Shah (2023) analyzed the food security in Pakistan. Timmer (2000) mentioned the macro dimensions of food security. Human Development Report (1994) highlighted the new dimensions of human security.

Pakistan

Pakistan, a country grappling with economic challenges, political uncertainty, cultural complexities, religious divisions, and an unstable democratic system, is categorized as a low-income developing country (Nandy, 2019). The agricultural sector, which serves as the backbone of the country, caters to the needs of Pakistan's rapidly growing population by providing food and fiber. Although the population growth rate has decreased from over 3% in the 1980s to 2.09% in 2009–2010, it is still considered relatively high. Despite a significant fivefold increase in wheat production during the same period, Pakistan remains a modest importer of this essential food crop. Bridging the gap between food supply and demand necessitates substantial efforts in terms of population control and technological advancements. The Pakistani government has been confronted with various challenges, including the rising inflation rate and widespread reports of a food crisis. The issue of food insecurity in Pakistan arises from multiple factors, such as rampant corruption, natural disasters, militarization, inadequate implementation of the country's food policy, and political instability. Moreover, Pakistan's reputation has been suffered globally due to its designation as a terrorist state. Consequently, receiving financial assistance from the international community, donor nations, and international financing organizations becomes essential.

Reasons Behind Food Insecurity

- Population displacement is a common consequence of disputes, leading to hindered access and disruption of agricultural productivity. Pakistan's economy has been severely affected by the multitude of crises caused by conflicts between various ethnic and religious groups.
- Economic disturbances, including joblessness, rising prices, and volatile economies, have significantly impacted the state of food security, especially in impoverished countries such as Pakistan. The repercussions of COVID-19 have further intensified the challenges faced by global food security, amplifying existing vulnerabilities that were already causing disruptions in food production, supply chains, and market accessibility in countries like Pakistan.
- Climate-related disasters such as droughts, floods, and storms have played a substantial role in exacerbating food insecurity and presenting a persistent threat to global agricultural systems, including those in Pakistan.
- The primary cause of the country's current food insecurity in 2022 can be attributed to the devastating floods that occurred in Pakistan. In 2022, the Government of Pakistan accused India of discharging maximum water to Pakistan , which caused an unprecedented flood situation. The "Indus Water Treaty" between the two countries controls water distribution between the two countries (Nandy, 2023a). These floods not only destroyed agriculture, crops, and livestock but also made it challenging for people to access a nutritious diet and an adequate amount of food. Additionally, Pakistan's food security situation has deteriorated further in the past year due to the rising food costs, declining food quality, and an increase in food consumption due to the impact of the flooding. Before the catastrophic monsoon floods that struck Pakistan in June and August of 2022, the yields of the "Rabi" wheat crop were affected by

heat waves in March and April of the same year. Furthermore, the scarcity of irrigation water and fertilizers further contributed to reduced animal production in most districts.

- Pakistan experienced a severe financial crisis in 2023, which led to a food crisis. In the fiscal year (FY) 2023–2024, the Pakistani government spent a substantial amount of money despite the extraordinary financial crisis. The government allocated 280 billion rupees for the military's development initiatives (Ansari, 2023).
- Pakistan has experienced a tumultuous period since the beginning of 2022, characterized by a severe financial crisis and ongoing political instability. According to the International Monetary Fund (IMF), the government's debt to its creditors has surpassed $77 billion. As a result of a deteriorating balance-of-payment situation, which has led to a decline in foreign reserves, it is expected that this substantial amount will be repaid over the course of the next three years (Hussain, 2023).
- The government of Pakistan suffered an acute financial crisis in 2023; in addition, the foreign exchange reserve dropped to its lowest level (Anees, 2023).

Food Crisis During COVID-19

Pakistan is the Islamic Republic according to its constitution, but in reality, it is a multireligious nation with Islamic factionalism (Nandy, 2017). When the entire world has been fighting COVID-19 together, then in Pakistan, governmental discrimination has been found in treating religious minorities. Religious intolerance has been manifested in Sunni Muslim-dominated Pakistan. The coronavirus has been referred to as the "Shia virus" to put more social pressure on the socially cornered minority Shia community (Nandy, 2022). During the pandemic, an allegation has been raised against the Sunni-owned ration shop for pressurizing Christians to recite the Quran for getting ration. The same attitudes were displayed against Hindus. Some government distribution centers had denied giving rations to the Hindus despite showing their national identity card. This is a hard fact in Pakistan, where the state agencies or allied agencies often do discriminatory acts toward religious minorities. The governmental policy of exclusion has been implied in society also. In many places, relief was distributed by private foundations and trusts. These organizations often deny assisting non-Muslim people to provide aid (Mirza, 2020)

The Domestic and International Response Toward Food Insecurity of Pakistan

According to "Vision 2030" of the Planning Commission of Pakistan, around half of the nation's population suffers from malnutrition, with the most vulnerable groups being women, children, and the elderly who belong to the lowest economic strata. To combat the issue of food insecurity, the Pakistani government has implemented several legislative measures and strategic actions. These include programs to boost agricultural production, import food, implement poverty reduction initiatives, enhance nutrition, and provide social safety nets. In 2022, Pakistan had terrible floods and heat waves, thereby ranking 8th out of 10 countries that are susceptible to natural disasters according to the Global Climate Risk Index. The Pakistani Federal Government implemented prompt and efficient policy measures to mitigate the socioeconomic consequences and promote economic recovery, even in the face of these unfavorable variables. The Federal Government additionally committed Rs. 70 billion for flood disaster aid.

Since the beginning of 2020, Pakistan has encountered a series of adversities exacerbating the nation's humanitarian requirements, especially in the provinces of Balochistan, Khyber Pakhtunkhwa (KPk), and Sindh. These adversities encompassed the invasion of desert locusts, monsoon flooding, harsh winter conditions, the COVID-19 pandemic, and the lingering consequences of the 2018–2019 drought. Approximately 11 million individuals in Pakistan were affected by these events. Despite the imminent threat of a severe food crisis caused by flooding, the situation could potentially deteriorate further. The initial formulation of a food security policy was undertaken by the province of Khyber Pakhtunkhwa. The government introduced the Kisan Card, which aimed to provide agricultural product subsidies. This program was a pioneering initiative of its kind, specifically designed to assist farmers in their endeavors (Dawn, 2022).

Prior to being self-dependent in producing food, Pakistan received food grain from the United States. In 2010, Pakistan was severely affected by floods, which consequently caused widespread damage to the economy. Apart from acute financial crisis, the food crisis was evident because of hampering irrigation, transport, and power sectors. The ADB sanctioned US $ 10 billion for assisting the Pakistani government to overcome the situation (Ahmed, 2011). The Government of Pakistan and community leaders receive support from the Bureau for Humanitarian Assistance (BHA) of the United States Agency for International Development (USAID) to effectively prepare for and mitigate the impact of natural disasters. This assistance involved the provision of cash aid, which aimed to restore the livelihoods of the affected communities and enhance their resilience to future shocks (USAID, 2021). Amid the devastating floods and the resulting disruptions to global supply chains, the ADB has approved a financial assistance package amounting to $1.5 billion. This funding aims to support the Government of Pakistan in its efforts to enhance food security, provide social protection, and create employment opportunities for its people.

Afghanistan

Afghanistan's limited foreign investments can be attributed to the resistance of its historic social structure to globalization and liberal economic policies. This has resulted in a severe crisis in the domestic economy. The lack of an education program by the government has contributed to the country's extreme underdevelopment. A consequence of this underdevelopment is the lack of employment opportunities for teenagers, leading some to join terrorist organizations. The Taliban's rise to power in 1996 further exacerbated the suffering of Afghanistan, as their violent activities undermined democratic institutions. During their rule, women were not allowed to work or pursue education abroad, and numerous journalists and pro-democracy campaigners were ruthlessly killed. The mistreatment of women by the Taliban until 2001 had a significant detrimental impact on their overall human rights, access to food, and well-being.

Food insecurity and malnutrition are some of the most serious manifestations of protracted crises, which disrupt livelihoods and food systems. Over half of the Afghan population, approximately 16 million people, lives below the poverty line, a sharp increase in the past decade. Rural poverty remains consistently higher than urban poverty, although the deterioration in welfare has become more widespread across the country. The high poverty rates represent the combined effect of stagnating economic growth, increasing demographic pressures, and a deteriorating security situation in the context of an already impoverished economy and society where human capital and livelihoods have been eroded by decades of conflict and instability. Food insecurity has increased in line with poverty, with 13 million Afghans between 2016 and 2017. In 2017–2018, Afghanistan experienced a drought that affected 22 of its 34 provinces, exacerbating food insecurity, triggering internal displacement of people, and adversely affecting investments in the 2018–2019 agricultural season (The World Bank, 2019, pp. 1–5). Changing climatic conditions, a growing population, and other environmental stressors will likely continue to have a significant impact on food security going forward. The calorie intake of the Afghan population is not only generally inadequate in quantity; it is also qualitatively poor. Overall, half of the population consumes food poorly or insufficiently , with the number being highest among rural residents.

Reasons for Food Insecurity in Afghanistan

Afghanistan is a war-torn state that is facing numerous problems related to human security (Nandy, 2019). The following are several reasons behind the food insecurity in Afghanistan:

- A burgeoning group of individuals facing hunger is emerging due to the repercussions of unemployment, financial constraints, and escalating prices. In Afghanistan, a staggering 13.8 million people are grappling with insufficient access to nourishment. The nation finds itself teetering on the brink of an economic collapse, exacerbated by the local currency plummeting to unprecedented lows and the relentless surge in food expenses.
- Afghanistan has been plagued by devastating droughts for the past four decades, leading to severe food shortages. Stretched water resources and the difficulties facing agriculture three years after the Taliban gained power in Afghanistan during the withdrawal of foreign soldiers are among the administration's main concerns. Humanitarian aid has decreased significantly this

year, and diplomats and charity workers report that there is no development support available to address the issue given that no foreign country has formally recognized the Taliban. Within the nation of around 42 million people, the UN World Food Programme reports that 15.3 million people experience severe food insecurity (Reuters, 2023).

- Due to prolonged conflicts, Afghanistan lacks the necessary resources and infrastructure to effectively manage risks such as food crisis.
- The economic turmoil following the Taliban's assumption of power in August 2021 has left many individuals unable to afford the rising costs of food. The Taliban's ascension to power did not cause the problems mentioned. According to the United Nations, approximately 681,600 people were displaced in Afghanistan due to conflicts between January and October 31, 2021. The dire situation in Afghanistan, characterized by a prolonged drought, violence, and economic collapse, has resulted in an alarming 55% of the population or an estimated 22.8 million individuals, facing severe food insecurity at crisis or emergency levels (Mil, 2021). This predicament predates the Taliban's takeover and has been exacerbated by the aforementioned factors.
- The consequences of climate change have been wide-ranging, but its impact on hunger has been especially severe. In Afghanistan, a significant portion of the population, specifically two out of every five individuals, face severe food insecurity. The agricultural sector plays a crucial role in providing both sustenance and income for 80% of families. Unfortunately, natural disasters, especially the recent droughts, have had detrimental effects on farmers, leading to a decrease in their food production and revenue. To address this issue, World Vision Afghanistan is actively working to enhance farmers' capabilities in adopting sustainable and climate-resilient agricultural practices. Additionally, they are focused on restoring productive assets that can withstand the frequent weather-related shocks experienced in the region. By implementing these measures, World Vision Afghanistan aims to empower households and communities, enabling them to become more self-reliant and secure in terms of food and livelihood.

COVID-19 and Food Insecurity

The food insecurity in Afghanistan has been exacerbated by the COVID-19 pandemic, with notable levels of blame being attributed to it. Additionally, the pervasive corruption within all branches of the government has further contributed to the country's food crisis. The presence of widespread corruption among politicians and officials has only served to worsen the situation of food insecurity in Afghanistan (Nandy, 2023b).

Food Insecurity under the Taliban Regime

Insufficient food security has resulted in acute malnutrition affecting a significant number of Afghan children, reaching a staggering figure of three million (Cost of

War, 2022). The dire situation is further exacerbated by ethnic conflicts and frequent natural disasters, leading to widespread starvation in numerous provinces across Afghanistan. A study conducted by Cost of War reveals that half of the Afghan population earns less than $1.90 per day on average. The US government allocated $36.07 billion from the US Treasury to address these pressing issues regarding human security in Afghanistan (Cost of War, 2022). After the Taliban assumed power in August 2021, concerns have arisen regarding the state of human security in Afghanistan. The installation of the Taliban regime quickly revealed the fragility of its administration. Tragically, numerous Afghan lives have been lost, countless individuals have been displaced from their homes, and a significant number have sought refuge in other nations. Moreover, the severe food crisis has further jeopardized the security of the population. Regrettably, there are no notable initiatives within Afghanistan that offer a viable solution to this dire predicament, as the Taliban administration appears to lack both empathy and a sense of responsibility toward its citizens. Due to finances, Afghans are unable to purchase essential goods, food, and medicines. Since August 2021, 9/10 households have continually faced insufficient food consumption (WFP, 2022). In Afghanistan, hunger has become alarming.

There is a complete disruption of the fundamental elements of human security, including life, livelihood, access to water, food security, shelter, and the public health system. Because the country's entire healthcare and educational systems were destroyed, the Afghan people have been denied access to basic medical care and quality education (EUAA, 2023, p. 10). After the Taliban took control of the government in August 2021, the foreign aid-dependent Afghan economy suffered a setback. Since August 2021, approximately half of Afghanistan's population has been grappling with hunger, with women and girls being disproportionately affected by this dire food crisis (International Crisis Group, 2023). In 2022, Afghanistan's agriculture industry faced a severe crisis due to a scarcity of fertilizers, water, seeds, and other essential resources, thereby highlighting the significance of human security (Nandy, 2024). The country is currently grappling with a severe food crisis as a result of the ongoing drought. Moreover, the restrictive measures imposed by the United States of America on the Central Bank of Afghanistan have disrupted crucial banking services, leading to the collapse of the nation's economy. Consequently, the Taliban government finds itself ill-equipped to address the multifaceted challenges posed by human security (World Report, 2023).

The Domestic and International Response Toward Food Insecurity in Afghanistan

The immediate lifting of restrictions on humanitarian supplies is imperative for the international community to address the worsening hunger crisis in Afghanistan. To ensure the long-term food security of the nation, it is crucial to enhance and fortify local agricultural systems and capabilities. From July 2021 to February 2022, India adopted a cordial approach toward Afghanistan, extending considerable aid in the realms of development and humanitarian assistance after the fall of the Taliban regime. In the aftermath of the Taliban's collapse, Afghanistan received noteworthy humanitarian support, with India playing a pivotal role by contributing a significant sum of $100 million toward the enhancement of infrastructure, healthcare facilities, and energy resources. Moreover, India's assistance encompassed various aspects of human security, healthcare, and education, with a particular focus on empowering women and children. In June 2023, India implemented a lenient approach to assist Afghanistan in its dire food predicament. The grain shipment was facilitated by Iran's Chabahar port (Parashar, 2023).

Results of the Study

On average, Pakistani households allocate 50.8% of their monthly income toward food expenses. This significant proportion renders them more vulnerable to various shocks, especially the escalation of food prices. The situation is further exacerbated by the adverse impacts of climate change and the displacement of populations. In addition to the prevailing economic instability, which has resulted in the prolonged storage of food and medicine at ports, the nation is currently grappling with the aftermath of devastating floods that took place during the summer season. Government authorities have reported that these floods have severely damaged more than 80% of the country's crops. Furthermore, the State Bank of Pakistan has recently disclosed that Pakistan's foreign exchange reserves have significantly declined to a critical level of $4.3 billion (Shah, 2023). The food crisis in Pakistan is a result of various political, natural, and economic factors. According to the National Nutrition Survey conducted in 2018, 36.9% of the population was found to be food insecure. The impending food crisis can be attributed to unpredictable and abnormal climatic conditions, which have led to catastrophic consequences and extensive damage to agricultural lands across the country (Ullah, 2022).

The study shows that Afghanistan is a country with a lot of hazards. The human security of the country has been affected for several interconnected reasons, such as terrorism, wars, lack of industrialization, infantile lands, agricultural backwardness, poverty, internal displacement, natural hazards, and the absence of democratic government. Additionally, corruption, lack of national leadership, and negative perception of the donor countries toward the Taliban-led government pose a threat to food security. The access to healthcare for women has been significantly curtailed due to the Taliban's restrictions on their employment and mobility (Faiez, 2024). The inhuman and undemocratic activities of the Taliban administration caused the stopping of foreign aid to Western countries which increased food insecurity. In the years 2017, 2018, and 2019, there were notable occurrences of floods and droughts. The pandemic not only caused food prices to escalate but also resulted in widespread unemployment, income losses, and disruptions in food supply networks. Consequently, a considerable number of Afghans are now facing difficulties in affording essential food items such as wheat and sugar, which have experienced price hikes of up to 23% (Lutz & Kurtzer, 2021).

Policy Recommendations

Based on the study, some policy recommendations can be offered to Pakistan and Afghanistan. For Pakistan – –

First, an adaptive safety net program has been developed to specifically address various vulnerabilities and consider the specific needs of different geographic regions, as well as account for seasonal fluctuations in hunger and livelihoods, along with other relevant factors (The World Bank, 2019, pp. 1–5). Second, following the expansion of safety nets, priority can be given to improving tertiary highways and implementing spatial targeting in areas that are either inaccessible or persistently neglected by markets. Third, in the medium term, efforts will be made to broaden the sources of income for rural residents and invest in livestock development and enhancement of animal product quality. Fourth, intermediate-term initiatives have been introduced to promote better nutrition and care for marginalized populations. These initiatives include school feeding programs, provision of nutritional supplements for mothers and children, and health campaigns targeting expectant and new mothers. Fifth, the efforts of local and national nongovernmental organizations (NGOs) are essentially required. Sixth, the Afghan people demand long-term, sustainable solutions that can be attempted by international funding agencies, and donor states to resolve the acute food crisis. Seventh, the government should try to adopt a "self-reliant policy" to mitigate the food crisis. Eighth, the welfare policy and implementation are to be ensured.

Conclusion

According to an assessment conducted by the World Bank on inflation and development in Pakistan, inflation was primarily driven by a combination of factors. These included the rise in fuel and oil subsidies, as well as transportation challenges caused by floods and the conflict in Ukraine, which had a significant impact on the importation of essential grains. Currently, the country relies on foreign loans with exorbitant interest rates to sustain its financial stability. In the upcoming years, it is obligated to repay a substantial amount of 80 billion dollars (Anees, 2023). In 2023, amid Pakistan's severe financial crisis, the presence of the IMF representatives in Islamabad was of utmost importance. Pakistan's foreign exchange reserves were alarmingly low (Turak, 2023).

Afghan households' heavy reliance on self-production of food makes them vulnerable to food insecurity risks by reducing their ability to produce food as well as their ability to generate income. Poor households disproportionately rely on reducing food quality or quantity or on selling assets or property, further compromising food security and resilience to shocks. Food insecurity is exacerbated not only by climatic occurrences such as drought but also by a lack of growth in agricultural output, which combined with population growth, leads to unequal access to agricultural resources. The current drought thus exacerbates already existing severe food insecurity in Afghanistan. In the short term, humanitarian assistance can reduce the urgent need for food, but in the medium to long term, food security objectives have to be incorporated into national policies on poverty reduction strategies. Strategies aiming at reducing hunger and poverty need to consider impacts at the national, subnational, household, and individual levels.

References

Ahmad, M., & Farooq, U. (2010). The state of food security in Pakistan: Future challenges and coping strategies. *Pakistan Development Review*, *49*(4), 903–923.

Ahmed, M. (2011). An economic crisis state? In M. Lodhi (Ed.), *Pakistan: Beyond the crisis state* (pp. 169–200). Rupa Publications India Pvt. Ltd.

Anees, S. M. (2023, May). Pakistan's economic crisis: What went wrong? *The Diplomat*. https://thediplomat.com/2023/05/pakistans-economic-crisis-what-went-wrong/. Accessed on February 14, 2024.

Ansari, U. (2023, June 22). Pakistan unveils increased defense budget, IMF decries spending plan. *Defence News*. https://www.defensenews.com/global/asia-pacific/2023/06/22/pakistan-unveils-increased-defense-budget-imf-decries-spending-plan/. Accessed on February 13, 2024.

Asian Development Bank (ADB). (2022). $1.5 Billion ADB financing to promote social protection. *Food Security in Pakistan*. https://www.adb.org/news/1-5-billion-adb-financing-promote-social-protection-food-security-pakistan. Accessed on February 14, 2024.

Cost of War. (2022). *Afghanistan before and after 20 years of war (2001–2021)*. Watson Institute for International and Public Affairs, Brown University. https://watson.brown.edu/costsofwar/Afghanistanbeforeandafter20yearsofwar. Accessed on February 11, 2024.

Dawn. (2022, November 29). *Pragmatic steps are taken to ensure food security: CM*. https://www.dawn.com/news/1723660. Accessed on February 12, 2024.

European Union Agency for Asylum (EUAA). (2023, January). *Country guidance: Afghanistan*. https://euaa.europa.eu/sites/default/files/publications/2023-01/2023_Country_Guidance_Afghanistan_EN.pdf. Accessed on April 23, 2023.

Faiez, R. (2024, February 12). Foreign aid drops sharply as Taliban abuses jeopardize the Afghan health system, the group says. *Washington Post*. https://www.washingtonpost.com/world/2024/02/12/hrw-foreign-aid-afghan-health-taliban/c1f75c26-c96e-11ee-aa8e-1e5794a4b2d6_story.html. Accessed on February 14, 2024.

Human Development Report 1994. (1994). *New dimensions of human security*. https://hdr.undp.org/content/human-development-report-1994. Accessed on February 11, 2024.

Hussain, A. (2023, July 26). Why is Pakistan ranked 99th on the Global Hunger Index? *Al Jazeera*. https://www.aljazeera.com/news/2023/7/26/why-is-pakistan-ranked-99th-on-the-global-hunger-index. Accessed on February 14, 2024.

International Crisis Group. (2023, February 23). *Taliban restrictions on women's rights deepen Afghanistan's crisis*. https://www.crisisgroup.org/asia/south-asia/afghanistan/329-taliban-restrictions-womens-rights-deepen-afghanistans-crisis. Accessed on February 11, 2024.

Khan, M. A., & Shah, A. A. S. (2010). Food insecurity in Pakistan: Causes and policy response. *Journal of Agricultural and Environmental Ethics*, *24*, 493–509. https://doi.org/10.1007/s10806-010-9274-2

Khetran, B. S. (2023). *Key drivers of food insecurity in Pakistan* (pp. 1–6). Institute of Strategic Studies. https://issi.org.pk/wp-content/uploads/2023/07/IB_Khetran_July_5_2022.pdf. Accessed on February 9, 2024.

Lutz, J., & Kurtzer, J. (2021, September 27). What the Taliban takeover means for food security in Afghanistan. *Centre for Strategic and International Studies*. https://www.csis.org/analysis/what-taliban-takeover-means-food-security-afghanistan. Accessed on February 11, 2024.

Maxwell, S., & Frankenberger, R. T. (1992). *Household food security: Concepts, indicators, measurements: A technical review* (pp. 1–274). UNICEF.

Mil, D. (2021). Afghanistan: Food insecurity crisis. *Graphic News*. https://www.graphicnews.com/en/pages/41974/afghanistan-food-insecurity-crisis. Accessed on February 12, 2024.

Mirza, J. A. (2020, April 28). COVID-19 fans religious discrimination in Pakistan. *Diplomat*. https://thediplomat.com/2020/04/covid-19-fans-religious-discrimination-in-pakistan/. Accessed on February 12, 2024.

Nandy, D. (2017). *Understanding Pakistan*. Kunal Books.

Nandy, D. (2019). Democracy, civil society and human security in a war torn state: A case study of Afghanistan. In P. Pal (Ed.), *India in South Asia: Challenges and opportunities* (p. 64). Kunal Books.

Nandy, D. (2022). Managing COVID-19: A comparative study of Islamic Republics of South Asia. In U. Rana & J. Govender (Eds.), *Exploring the consequences of the COVID-19 pandemic: Social, cultural, economic, and psychological insights and perspectives* (pp. 1–13). Apple Academic Press (Taylor and Francis Group).

Nandy, D. (2023a). Rethinking Indus Water Treaty between India and Pakistan: People, politics, and policy. In C. Ploberger (Ed.), *River basins and international relations: Cooperation, conflict and sub-regional approaches* (pp. 107–123). Routledge.

Nandy, D. (2023b). The culture of corruption in Pakistan and Afghanistan: Impacts on socio-economic profiles. In R. Chandra Das (Ed.), *Social sector spending, governance and economic development: Perspectives from across the world* (pp. 126–139). Routledge.

Nandy, D. (2024). Human security and livelihood in Afghanistan. In D. Nandy & D. Majee (Eds.), *Human security in Asia: Interrogating state, society, and policy* (pp. 229–252). Palgrave Macmillan.

Parashar, S. (2023, June 11). Amid Taliban outreach, India to boost food aid. *The Times of India*. https://timesofindia.indiatimes.com/india/amid-taliban-outreach-india-to-boost-food-aid/articleshow/100905720.cms. Accessed on February 11, 2023.

Peng, W., & Berry, E. M. (2019). The concept of food security. In P. Ferranti, E. M. Berry, & J. R. Anderson (Eds.), *Encyclopedia of food security and sustainability* (Vol. 2, pp. 1–7). Elsevier.

Pinstrup-Andersen, P. (2009). Food security: Definition and measurement. *Food Security*, *1*, 5–7. https://doi.org/10.1007/s12571-008-0002-y

Reuters. (2023, August 15). *Prolonged drought deepens Afghanistan's humanitarian crisis.* https://www.reuters.com/world/asia-pacific/prolonged-drought-deepens-afghanistans-humanitarian-crisis-2023-08-11/. Accessed on February 12, 2024.

Shah, S. (2023, January 13). Why Pakistan is facing a growing food crisis. *Time.* https://time.com/6247300/pakistan-food-crisis-climate-change-inflation/. Accessed on February 14, 2024.

The World Bank. (2019). *Afghanistan and development brief: Food insecurity in Afghanistan Five key policy-relevant findings about food insecurity* https://documents1.worldbank.org/curated/en/690371563979448707/pdf/Afghanistan-Development-Brief-Food-Insecurity-in-Afghanistan-Five-Key-Policy-Relevant-Findings-about-Food-Insecurity.pdf. Accessed on February 11, 2024.

Timmer, C. P. (2000). The macro dimensions of food security: Economic growth, equitable distribution, and food price stability. *Food Policy*, *25*, 283–295.

Turak, N. (2023, February 3). Blackouts, currency dives, and corruption: Pakistan's economy is on the brink of collapse. *CNBC.* https://www.cnbc.com/2023/02/03/blackouts-currency-dives-and-corruption-pakistans-economy-is-on-the-brink-of-collapse-.html. Accessed on February 14, 2024.

Ullah, S. M. (2022, December 22). Looming food crisis in Pakistan. *The Nation.* https://www.nation.com.pk/18-Dec-2022/looming-food-crisis-in-pakistan. Accessed on February 14, 2024.

USAID. (2021). *Bureau for humanitarian assistance: Pakistan assistance overview.* https://www.usaid.gov/sites/default/files/2022-05/USAIDBHA_Pakistan_Assistance_Overview_-_July_2021.pdf. Accessed on February 13, 2024.

World Food Programme (WFP). (2022). *Afghanistan food security update round.* https://www.wfp.org/emergencies/afghanistan-emergency. Accessed on January 23, 2023.

World Report. (2023). *Afghanistan events of 2022.* Human Rights Watch. https://www.hrw.org/world-report/2023/country-chapters/afghanistan. Accessed on April 21, 2023.

Chapter 15

Understanding the Linkage Between Household Poverty and Food Security: Insights From the Indian State of Odisha

Arpita Das[a] *and Bidur Paria*[b]

[a]Indian Institute of Technology Guwahati (IIT Guwahati), India
[b]B. R. Ambedkar School of Economics University, India

Abstract

In India, more than 21.9% of the total population is below the poverty line, which compelled the government to initiate several policies for alleviating poverty and food insecurity. It can be argued that often non-poor households may become food insecure due to their irrational expenditure behaviour. The widespread consumerism and commercialism in rural India as a result of globalization often drive rural households to spend more on non-food items than food items. As a result, it is frequently observed that non-poor farmers excluded from the social security programs are highly malnourished as a result of food insecurity, which pushes them back into the poverty trap. This study reviews critically the methodologies applied to measure household poverty and food insecurity and examine the relationship between the two using primary data collected from two villages of the Balasore district of Odisha whose major livelihood relies on agriculture using the m-logit model for data analysis. Results show even households living above the poverty line suffer from the problem of food insecurity. About 20.5% of the non-poor are food insecure in the study villages. The study also finds a location-specific spatial autocorrelation of Multidimensional Poverty Index (MPI) and headcount ratio among the districts of the state using Moran's I. Hence, there is a need for properly targeted policy interventions and to revise the methodologies applied to measure household poverty and identify beneficiaries so that non poor households are not deprived of food security programmes.

Keywords: Food security; Moran's I; non-poor; poverty; agriculture; Odisha

Growth, Poverty and Developmental Aspects of Agriculture, 225–247

doi:10.1108/978-1-83608-076-320241015

Introduction

The concept of food security has evolved significantly over the past three decades in response to changes in official policy perspectives (Clay, 2002; Heidhues et al., 2004). Initially stemming from discussions during the World Food Conference in the mid-1970s, the focus was primarily on ensuring the availability and stability of essential food supplies at national and global levels (World Food Conference, 1974). By 1983, the emphasis shifted towards food access, considering the balance between demand and supply in food security assessments. The definition was further broadened with the 1986 World Bank Report on Poverty and Hunger, which introduced the distinction between chronic and transitory food insecurity. Chronic food insecurity was associated with persistent low incomes and structural poverty, while transitory food insecurity referred to temporary pressures resulting from economic downturns, natural calamities or conflicts. Moreover, Sen's theory of famine in 1982 underscored the role of individual entitlements in accessing food, taking into account factors like production, trade, labour and resources obtained through transfers. The widely accepted definition adopted at the 1996 World Food Summit emphasized the multifaceted nature of food security, encompassing aspects such as food access, availability, utilization and stability. In recent years, increasing attention has been given to the ethical and human rights dimensions of food security, advocating for the Right to Food. Finally, in 2004, an Intergovernmental Working Group under the Food and Agriculture Organization (FAO) Council developed a set of voluntary guidelines aimed at supporting the progressive realization of the right to adequate food within the framework of national food security (FAO, 2006).

According to FAO (2006), food security entails four main components: the availability of nutritious food, the ability to afford and access food both economically and physically, the efficient utilization of food and the sustainability of these factors (Irfeey et al., 2023). Various methodologies are commonly employed to assess food insecurity, including the Household Food Insecurity Access Scale (HFIAS) (Castell et al., 2015), Household Food Security Survey Module (HFSSM) (Frongillo, 1999; Radimer et al., 1990), Food Insecurity Experience Scale (FIES) (Ballard et al., 2014) and Radimer/Cornell measures of hunger and food insecurity (Kendall et al., 1995), among others. These approaches typically involve categorization, scaling and scoring based on surveys and interviews. Similarly, multidimensional poverty encompasses various dimensions, including health, political, social, economic and environmental factors (Gweshengwe & Hassan, 2020). These dimensions involve considerations such as having income levels below the minimum wage or poverty line (Banerjee, 2016), lacking adequate resources to meet basic needs (SIDA, 2015), facing unemployment or working in low-paying jobs (Hulme & McKay, 2013, pp. 187–214), experiencing limited social connections (Raphael, 2011) and encountering barriers to accessing infrastructure, education, clean water and electricity (Narayan et al., 2000). The assessment of poverty primarily relies on consumption expenditure data utilizing a monetary metric approach (Chauhan et al., 2016; Das et al., 2022). These data are compared against a predefined poverty threshold established for the specific region or area.

In the pursuit of food security, there's a common assumption that ensuring an abundant food supply is paramount. However, Amartya Sen's research highlights that merely having sufficient food available does not guarantee food security, as evidenced by historical famines like the Great Bengal famine and the Sahelian famines, which occurred despite adequate food availability. The crux of the issue lies in factors affecting access, such as challenges related to income distribution and affordability. Policies that prioritize food availability over access can pose threats to food security (Pérez-Escamilla, 2017). Moreover, policies that restrict small-scale farmers from cultivating cash crops or compel them to prioritize food crop production, even when cash crops could yield higher incomes, may undermine their food security by reducing their real incomes (Martin, 2010). The idea that falling prices benefit consumers fails to consider the impact on rural producers and those reliant on them for employment, as inflation can negatively affect their incomes. Observing a decline in the share of food expenditure may not necessarily indicate improved well-being, particularly in contexts with low standards of living like India. Such declines may occur when people are actually experiencing deteriorating conditions, as increased monetization and higher utility costs force individuals to allocate more funds to essential non-food items, such as fuel. Consequently, individuals may need to reduce their food expenditure to cover transportation, healthcare and other necessities, affecting primarily staple grains given their significant share of food expenditure (Patnaik, 2005).

In addition, government initiatives offering support for housing, utilities and nourishment, while helpful, still leave many individuals struggling financially. This situation contributes to the welfare trap. The welfare trap is a situation where government assistance programmes intended to alleviate poverty inadvertently contribute to its perpetuation. Despite government initiatives aimed at providing support for housing, utilities and nutrition, many individuals still struggle financially. Poverty traps can arise from various factors, including limited access to nutritious food, education, corrupt governance or the impacts of climate change. Ironically, welfare traps often result from policies intended to alleviate poverty, such as means-tested government assistance programmes. These programmes offer subsidies for essentials like housing, food and healthcare, but individuals lose access once their income exceeds a specific threshold, regardless of their financial stability. This perpetuates a cycle of poverty, discouraging individuals from seeking employment and stifling economic growth. While some propose eliminating government assistance programmes altogether, this approach is considered impractical (Bay, 2022).

Progress towards achieving Sustainable Development Goal (SDG), Targets 2.1 and 2.2, which aim to eliminate hunger, ensure year-round access to safe and nutritious food and eradicate malnutrition, was already faltering prior to the onset of the COVID-19 pandemic. The crisis exacerbated existing challenges, exposing vulnerabilities and inequalities within the global food system. Despite these setbacks, there are valuable lessons to be gleaned from these experiences, offering insights that can guide efforts to revitalize progress toward meeting these SDG targets. This global assessment serves as a diagnostic tool, elucidating the essential policies that should be implemented to address these pressing issues (FAO, IFAD, UNICEF, WFP, & WHO, 2021).

In India, despite substantial economic growth and transformations in recent decades, the improvements in financial well-being have not significantly alleviated food insecurity and malnutrition among the population (Saxena, 2018). Despite increased food production and availability, these advancements have not led to notable improvements in overall nutrition outcomes. The widening income gap in India has adverse effects on health outcomes (Narain et al., 2016), leading to increased attention on issues of food insecurity and nutrition, prompting research efforts to measure and understand these phenomena (McKay et al., 2023). Various factors contribute to food insecurity in India, including ineffective food policies that fail to address underlying issues, challenges posed by droughts and floods at various levels and national policies that prioritize large-scale farmers over small-scale ones. This neglects critical aspects such as infrastructure, food crop management, storage, land distribution, pricing and conservation (Oluoko-Odingo, 2011; Baral et al., 2021). Additionally, climate change has been identified as another contributing factor to food insecurity in the country (Ali & Erenstein, 2017).

Based on these concerns, this study primarily aims to measure household poverty and food insecurity and to find out the factors that impact food security in the coastal district of Balasore in the state of Odisha. The first section briefly introduces the problems and scenario of food security and poverty, followed by methodology, results and discussion and finally concludes.

Materials and Methods

Study Area Selection

The study employs a multi-stage random sampling technique. Initially, Odisha state is chosen in the first stage. Odisha is an agrarian state, with about 70% of its population engaged in this sector. Despite its pivotal role, agriculture contributes only around 26% to the Gross State Domestic Product (GSDP). The average landholding size of the state is 1.25 ha, 83% of small and marginal farmers. The state encompasses a cultivable area of about 64.09 lakh ha out of its total geographical expanse of 155.711 lakh ha, constituting 41.16%. Within this area, approximately 61.50 lakh ha are under active cultivation. Odisha's diverse geography led to its division into 10 agro-climatic zones based on factors like soil composition, humidity, elevation and rainfall patterns. With an average annual rainfall of 1,452 mm, predominantly during the monsoon months from June to September, the state's irrigation infrastructure caters to 27.63 lakh ha in Kharif and 13.31 lakh ha in Rabi seasons. Rice cultivation predominates in Odisha's agricultural landscape (Directorate of Horticulture, Govt. of Odisha). In the second stage, among 30 districts of the state, the Balasore district is randomly selected. Balasore, a coastal district, is predominantly agrarian and known for its cultivation of rice (Dhan), betel (Paan) and fish (Meen). However, the district faces challenges such as irregular income and variable productivity due to recurrent natural disasters. Balasore has approximately 2.50 lakh ha of cultivated land, with 2.10 lakh ha devoted to rice cultivation in the Kharif season and 1.07 lakh ha in Rabi, also including pulses, oilseeds and vegetables. In the third stage, two blocks, Jaleswar and Remuna, are randomly selected. Subsequently, in the fourth stage, two villages are chosen: Gopimohanpur from the Jaleswar block and Armala from the Remuna block. Finally, in the last stage, a total of 100 households are randomly selected, with 50 households chosen from each village, to assess their consumption and farming practices in the year 2022.

Sources of Data

The study employs both secondary and primary data. The secondary data on headcount ratio (HCR) and Multidimensional Poverty Index (MPI), for the period 2015–2016 and 2019–2021 are collected from 'A Progress Review 2023, National Multidimensional Poverty Index' and cropping intensity and irrigation intensity for the period 2015–2016, and 2018–2019 are collected from Odisha Agriculture Statistics. A multi-stage random sampling procedure is followed to collect the primary data. Out of the 12 blocks, the Jaleshwar and Remuna blocks are selected at random. We chose Gopimohanpur and Armala as our sample villages. Ultimately, a total of 100 respondents, 50 individuals from each town whose primary source of income is agriculture, were interviewed.

Methodology

Terminology Related to Poverty and Food Insecurity Measurement

Headcount ratio – It is the proportion of the population living below the poverty line.[1] It shows how many people are poor.

Poverty Gap – It illustrates the average shortfall of the population from the poverty line, providing insight into the intensity of poverty within the economy and indicating the level of impoverishment among the poor.

Square Poverty Gap – It is the square of the average poverty gap. It shows the severity of poverty among the population. By squaring the poverty gap for each household or individual, this metric assigns greater significance to those who experience substantial shortfalls below the poverty line compared to those who are relatively closer to it.

Food Insecurity – It represents the proportion of the population experiencing food insecurity from the perspective of their expenditure on food.[2]

---

[1]The poverty line for rural Odisha, adjusted for inflation, is set at Rs 1,558.04 for the year 2021, as determined by the Rangarajan poverty estimation report. This figure is based on monthly per capita consumption expenditure.

[2]The food security line for rural Odisha is estimated at Rs 888.08 based on MPCE on food expenditure determined by Rangarajan Committee report.

Food Insecurity Gap – This is determined by calculating the mean deviation of the population from the food security line, providing insight into the average shortfall of the population in meeting the criteria for food expenditure.

Square Food Insecurity Gap – It is the square of the average food insecurity gap. It shows the severity of food insecurity among the population.

Multinomial Logit Model

To determine the factors affecting food security and poverty of the sample households, multinomial logistic (m-logit) regression model is used (Aboaba et al., 2020). The multinomial logit model determines the probability that household i experiences one of j outcomes is given by:

$$P_{ij} = E\left(Y_i = \frac{j}{X_i}\right) = F(\alpha + \beta_i X_i), j = 1, 2, 3, 4$$

$$= \frac{1}{1 + \sum_{j=1}^{3} e^{z_i}} \text{ (where, } z_i = \alpha + \beta_i X_i + e_i)$$

where $P_{ij} = E\left(Y_i = \frac{j}{X_i}\right)$ is the probability of the reference category.

Differentiating the above equation, we get the marginal effects of the independent variables by the following equation:

$$\delta_j = \frac{\partial P_i}{\partial X_i} = P_j\left(\beta_j - \sum_{k=1}^{3} Pk\beta K\right) = P_j(\beta_j - \theta)$$

$$P_{ij} = \frac{e^{z_i}}{1 + \sum_{j=1}^{3} e^{z_i}}$$

P_{ij} is the probability of being in each of the groups compared to the reference/base category.

$P_{ij} = 1$, if the household is Food Insecure but Non-Poor (FINP); $0 =$ otherwise.

$P_{ij} = 2$, if the household is Food Secure but Poor (FSP); $0 =$ otherwise.

$P_{ij} = 3$, if the household is Food Insecure and Poor (FIP); $0 =$ otherwise.

Both Food Secure and Non-Poor (FSNP) are considered to be the base category to which other dependent variables are compared.

X_i are the independent variables.

X_1 is the age of the household's head (in years), X_2 is the square of the age of the household's head, X_3 is self-employment ($1 =$ if any member is self-employed; 0 otherwise), X_4 is regular employment ($1 =$ if any member is regularly employed; 0 otherwise), X_5 is the share of agricultural income to total income, X_6 is the per capita agricultural land, X_7 is the per capita Public Distribution System (PDS) consumption, e_i is the disturbance term.

Ordinary Least Squares (OLS) Regression

To examine the relationship and quantify the extent of association between the dependent variables, HCR (in percentage) and MPI and the independent variables, namely cropping intensity and irrigation intensity, OLS regression analysis is conducted. This method aims to minimize the sum of squares of estimated parameter errors, by following required theoretical assumptions. The OLS regression model can be expressed through the following equation:

$$y_i = \beta + \alpha_1 \mathrm{CI} + \alpha_2 \mathrm{II} + \epsilon$$

where y is the dependent variable ($i = 1, 2$; 1 for HCR, 2 for MPI), CI (Cropping Intensity) and II (Irrigation Intensity) are the independent variables, β is the intercept term, α_1 and α_2 are the respective slope coefficients of CI and II and ϵ is the error term.

Spatial Autocorrelation (Moran's I)

Spatial autocorrelation refers to the examination of how attributes and their spatial locations correlate. It is commonly assessed using Moran's I, which ranges from -1 to 1 and measures the degree of similarity between neighbouring locations. A positive Moran's I value indicates clustering of similar values, while a negative value suggests dispersion. When Moran's I is close to zero, it suggests no spatial pattern or correlation. This analysis helps understand the spatial distribution of attributes and identify any clustering or randomness within the dataset with an inference based on a null hypothesis of spatial randomness (Isazade et al., 2023). It is a cross-product statistic between a variable and its spatial lag, with the variable expressed in deviations from its mean. The tool also calculates a z-score and p-value to evaluate the significance of the model. The following equation calculates the Moran's I:

$$I = \frac{N \sum_i \sum_j w_{ij} \left(x_i - \overline{x}\right)\left(x_j - \overline{x}\right)}{W \sum_i \left(x_i - \overline{x}\right)^2}$$

where N is the number of observations, w_{ij} is the spatial weight matrix, x is the variable, $\overline{x}$ is the mean of the variable x and W is the sum of all spatial weights.

Results and Discussion

The study incorporates both secondary and primary data. Section 'Secondary Data Analysis' outlines the analysis of secondary data, while Section 'Primary Data Analysis' focuses on the analysis of primary data.

Secondary Data Analysis

Fig. 15.1 represents the intra-district variability of the HCR to measure poverty for the years 2015–2016 and 2019–2021 along with their respective Moran's I Index values for spatial autocorrelation assessment. Here, for intra-district variability and to know the intra-district disparities, we have considered the standard deviation from the mean HCR of Odisha, which stood at 29.34% for 2015–2016 and 15.68% for 2019–2021. Negative standard deviations indicate lower poverty and positive implies higher poverty comparatively. In Fig. 15.1A, coastal districts like Balasore, Jajpur, Kendrapara, Cuttack, Jagatsinghpur, Puri, Ganjam and central districts such as Jharsuguda, Sambalpur, Angul, Nayagarh and Khurda exhibit HCRs below a standard deviation of −0.5 from the state average.

Similarly, most western districts like Nuapara, Bolangir, Bargarh, Sonepur, Boudh and Sundargarh have headcount percentages within a standard deviation from −0.50 to 0.50%. Some of the northern districts, such as Mayurbhanj and Keonjhar, and southern districts, such as Kalahandi, Kandhamal, Rayagada and Gajapati, lie in the positive range of 0.50–1.5 percentage standard deviation. However, some northern districts like Mayurbhanj and Keonjhar and southern districts such as Kalahandi, Kandhamal, Rayagada and Gajapati fall within a positive range of 0.50–1.5 percentage standard deviation. Southern districts, particularly Malkangiri, Koraput and Nabarangpur, are the most affected, with higher HCRs compared to other areas. Fig. 15.1E shows an overall improvement in the poverty status of the state in as the state's average HCR has decreased to 15.68% in 2019–2021 from 29.34% in 2015. Coastal districts display the lowest headcount percentage as compared to other districts, while Malkangiri still remains a highly affected district. From Fig. 15.2A and C, it is also observed that the districts having similar HCRs are clustered together, which means there exists spatial autocorrelation, which is location-specific. In order to prove this, we have used Moran's I Index, which is shown in Fig. 15.2B and D for the years 2015–2016 and 2019–2021, respectively. The z-scores for the years 2015–2016 and 2019–2021 are 4.76 ($p = 0.000002, p < 0.05$) (Fig. 15.1B) and 3.98 ($p = 0.000068$, $p < 0.05$), respectively, with significant p-values (Fig. 15.1D). The positive significant z-scores indicate a tendency to cluster, indicating the location-specific nature of poverty in both years.

Turning to Fig. 15.2, a similar analysis is conducted considering the MPI for the years 2015–2016 and 2019–2021 (Fig. 15.2E). The state's average MPI for the years 2015–2016 and 2019–2021 are calculated to be 0.13 and 0.07, respectively. Similar to that of the HCR in Fig. 15.1, it is also observed that the coastal and central districts continue to demonstrate relatively lower MPI values, indicating improved multidimensional well-being. Conversely, districts in southern Odisha, such as Malkangiri and Koraput, have persistently high MPI values, indicative of deprivation across multiple dimensions. The spatial clustering of MPI values, as illustrated in Fig. 15.2B and D, further underscores the localized nature of poverty dynamics. Moran's I Index validates these observations, with significant positive z-scores ($z = 4.62$ ($p = 0.000004, p < 0.05$) for 2015–16 (Fig. 15.2B) and $z = 2.56$ ($p = 0.000367$), $p < 0.05$) for 2019–21 (Fig. 15.2D) affirming the presence

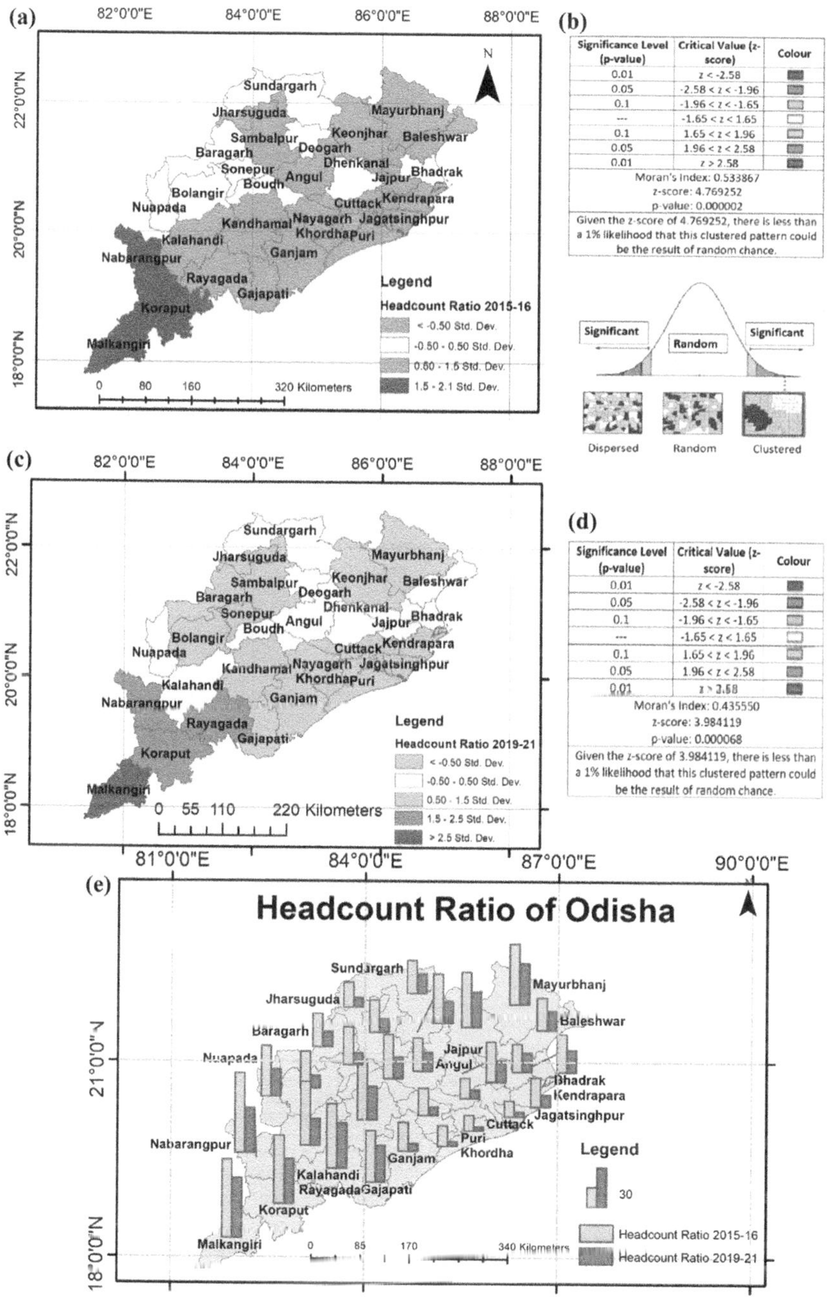

Fig. 15.1. Intra-district Variability of the Headcount Ratio to Measure Poverty. *Source:* India: A Progress Review 2023, National Multidimensional Poverty Index and authors' analysis using Arc Map. *Notes:* Fig. 15.1A - Standard deviation of headcount ratio of the year 2015–2016; Fig. 15.1B Moran's I for the year 2015–2016; Fig. 15.1C Standard deviation of headcount ratio of the year 2019–2021; Fig. 15.1D Moran's I for the year 2019–2021; Fig. 15.1E Comparison of headcount ratio between the years 2015–2016 and 2019–2021.

(a)

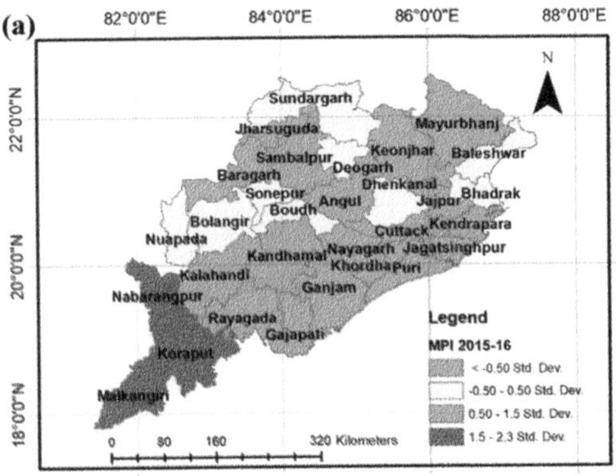

(b)

Significance Level (p-value)	Critical Value (z-score)	Colour
0.01	z < -2.58	
0.05	-2.58 < z < -1.96	
0.1	-1.96 < z < -1.65	
---	-1.65 < z < 1.65	
0.1	1.65 < z < 1.96	
0.05	1.96 < z < 2.58	
0.01	z > 2.58	
Moran's Index: 0.514054 z-score: 4.625930 p-value: 0.000004		
Given the z-score of 4.62593049419, there is less than a 1% likelihood that this clustered pattern could be the result of random chance.		

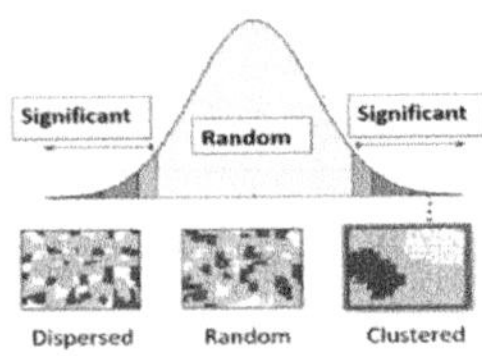

(c)

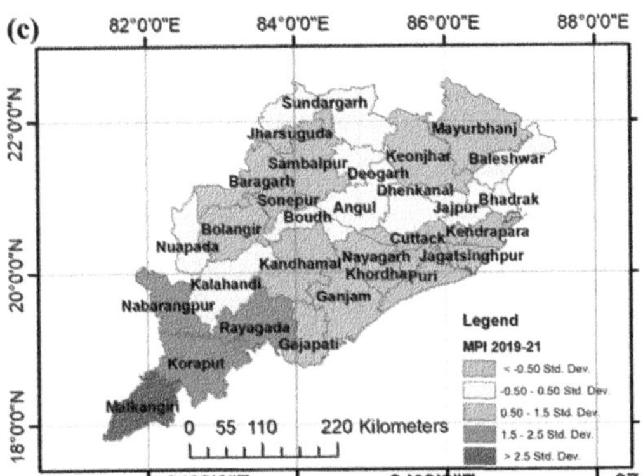

(d)

Significance Level (p-value)	Critical Value (z-score)	Colour
0.01	z < -2.58	
0.05	-2.58 < z < -1.96	
0.1	-1.96 < z < -1.65	
---	-1.65 < z < 1.65	
0.1	1.65 < z < 1.96	
0.05	1.96 < z < 2.58	
0.01	z > 2.58	
Moran's Index: 0.388555 z-score: 3.562979 p-value: 0.000004		
Given the z-score of3.56297902169, there is less than a 1% likelihood that this clustered pattern could be the result of random chance.		

(e)

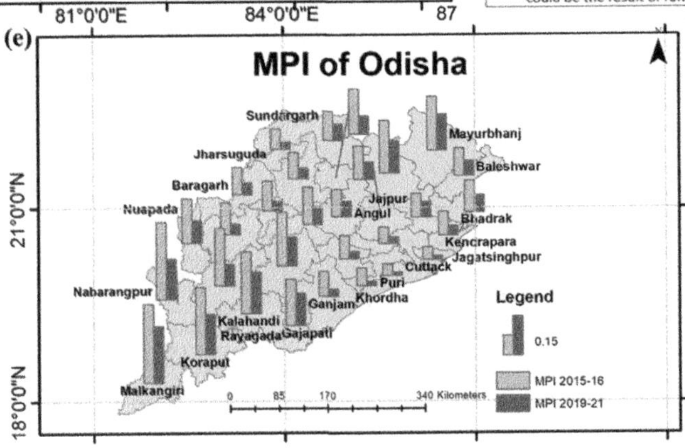

Fig. 15.2. Multidimensional Poverty MPI. *Source:* India: A Progress Review 2023, National Multidimensional Poverty Index and authors' analysis using ArcMap. *Notes:* Fig. 15.2A Standard deviation of MPI of the year 2015–2016; Fig. 15.2B Moran's index for the year 2015–2016; Fig. 15.2C Standard deviation of MPI of the year 2019–2021; Fig. 15.2D Moran's index for the year 2019–2021; Fig. 15.2E Comparison of MPI between the years 2015–2016 and 2019–2021.

of spatial clustering, This phenomenon emphasizes that poverty is not randomly distributed but rather exhibits spatial patterns, suggesting that factors influencing poverty are often location-specific.

The analyses infer that poverty rates are lowest in the coastal region, while they are highest in the southern and northern areas. Despite efforts to reduce poverty through initiatives like the Revised Long-Term Action Plan, Biju KBK Plan, Biju Kandhamal O Gajapati Yojana, Gopabandhu Gramin Yojna, Backward Region Grant Fund, Western Odisha Development Council and various other schemes, for the later regions, still the state's development remains uneven. Despite abundant natural resources like forests and minerals in these regions, they remain underutilized, with few industries established. Furthermore, agriculture in these regions relies heavily on monsoon rains, with inadequate measures taken to improve irrigation and agricultural marketing. The farmer suicides, along with migrations in these regions, are the reasons due to crop failures and unavailability of income opportunities (Pradhan, 2017). On the other hand, the coastal regions have more advanced agricultural development compared to other parts of the state. The advanced agricultural development in these areas is facilitated by their flat, fertile and alluvial soil, wider irrigation coverage, increased fertilizer usage and relatively higher rainfall levels (Patra, 2014).

Fig. 15.3 displays the results of the OLS regression analysis. In this analysis, the dependent variables are the HCR and the MPI, while the independent variables are cropping intensity and irrigation intensity. The figure illustrates that both cropping intensity and irrigation intensity have a negative relationship with HCR ($R^2 = 0.30$ for 2015 (Fig. 15.3A, A1, A2); $R^2 = 0.30$ for 2019 (Fig. 15.3B, B1, B2)) and MPI ($R^2 = 0.28$ for 2015 (Fig. 15.3D, D1, D2); $R^2 = 0.29$ for 2019 (Fig. 15.3E, E1, E2)). However, the relationship of both of the dependent variables with irrigation intensity is statistically significant (10% level of significance), indicating that an increase in irrigation intensity is associated with a decrease in poverty measures. The negative relationship between irrigation intensity and poverty measures suggests that higher levels of irrigation intensity are associated with lower levels of poverty, as indicated by the HCR and MPI. Irrigation allows farmers to cultivate crops more by providing a consistent water supply, especially during dry seasons. This improved access to water through irrigation enables farmers to diversify their crop production, making households less reliant on a single crop. Additionally, irrigation systems help buffer against the adverse effects of climate variability, such as droughts or erratic rainfall patterns. Thus, maintaining and improving their major source of livelihood (agriculture), irrigation plays a vital role in farming households of a state like Odisha.

Primary Data Analysis

Consumption-expenditure-based Poverty Estimation

We've evaluated the poverty status of the sampled households by considering their per capita monthly expenditure (MPCE), which includes both food and

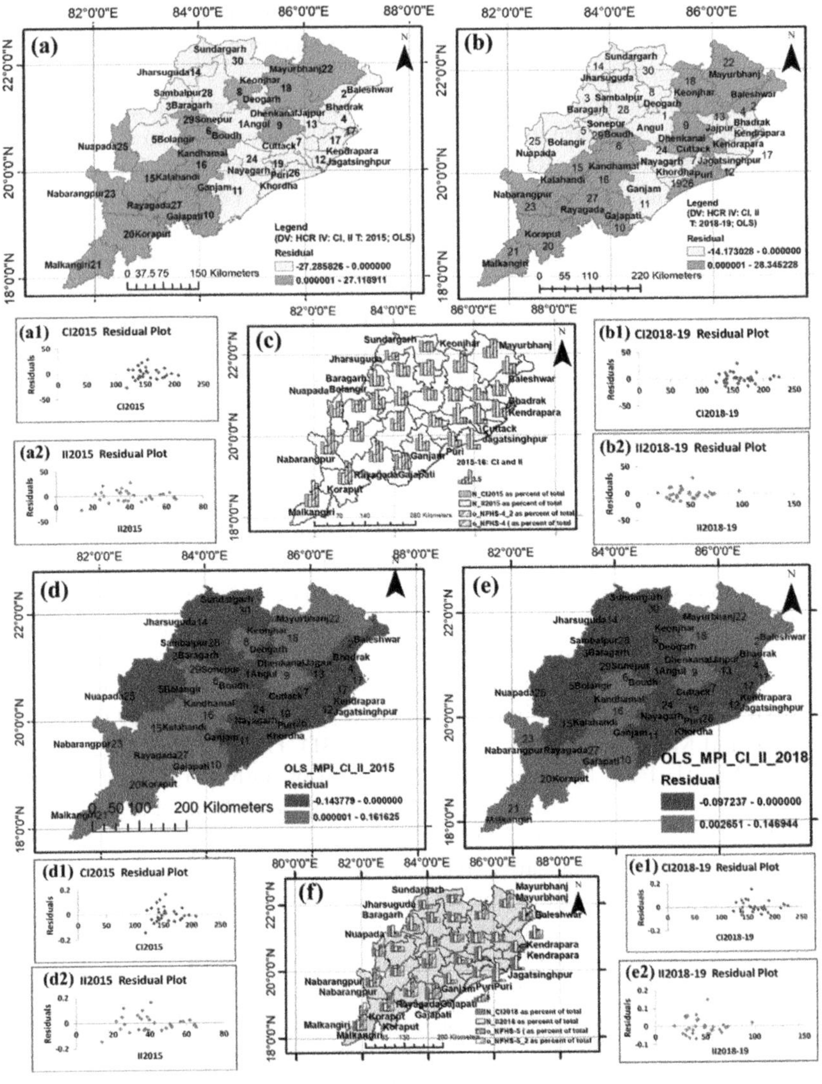

Fig. 15.3. Results of the OLS Regression Analysis. *Source:* India: A Progress Review 2023, National Multidimensional Poverty Index, Agriculture Statistics of Odisha 2015, 2018 and authors' analysis using ArcMap. *Notes:* Fig. 15.3A Residual variations across districts of headcount ratio as dependent variable and CI, II as independent variables for the year 2015–2016; Fig. 15.3A1 and A2 Residual plots for headcount against cropping intensity and irrigation intensity, respectively, for the year 2015–2016; Fig. 15.3B Residual variations across districts of headcount ratio as dependent variable and CI, II as independent variables for the year

non-food expenses (Das et al., 2022). Food expenditure involves purchases like grains, vegetables, eggs, meat and fish, while non-food spending encompasses expenses such as oil, entertainment, electricity, telephone bills, education, healthcare and contributions to social causes. The poverty line acts as the benchmark expenditure level that separates individuals into poor and non-poor categories. Employing this method allows us to examine the distribution of household incomes relative to the poverty threshold.

Table 15.1 illustrates three poverty indicators: the HCR, which depicts the incidence of poverty; the poverty gap, which measures the depth of poverty; and the square poverty gap, which signifies the severity of poverty among the sample households. Using the inflation-adjusted per capita monthly expenditure (MPCE) of Rs 1,558.04 as the poverty line for rural Odisha, the total HCR for the sample households is calculated at 26.12%, indicating that 26.12% of households fall below the Rs 1,558.04 threshold. Similarly, the poverty gap is determined to be 2.97, representing the average income shortfall of individuals or households below the poverty line. The square of the poverty gap, known as the severity of poverty, provides insight into the inequality among the population below the poverty line, with a calculated severity of 0.54.

Table 15.1 presents the poverty estimation for sample households categorized by their employment status and per capita landholding size. The households are classified into three main groups: casual labourers (CL), regular employees (RE) and self-employed (SE). CL typically work in various sectors such as hotels, shops, tailoring and manual labour, often without consistent year-round employment and based on job availability. Regular employees, on the other hand, work in nearby factories, petrol pumps or engage in teaching, usually on a full-time or part-time basis with regular shifts and pay. The self-employed category encompasses individuals who operate their own farming or non-farming activities independently, including agriculture farmers and those engaged in retail shops or tailoring.

2019–2021; Fig. 15.3B1 and B2 Residual plots for headcount against cropping intensity and irrigation intensity, respectively, for the year 2019–2021; Fig. 15.3C Comparison using bar diagram between CI, II, HCR and MPI in the years 2015–2016; Fig. 15.3D Residual variations across districts of MPI as dependent variable and CI, II as independent variables for the year 2015–2016; Fig. 15.3D1 and D2 Residual plots for MPI against cropping intensity and irrigation intensity, respectively for the year 2015–2016; Fig. 15.3E Residual variations across districts of MPI as dependent variable and CI, II as independent variables for the year 2019–2021; Fig. 15.3E1 and E2 Residual plots for MPI against cropping intensity and irrigation intensity, respectively for the year 2019–2021; Fig. 15.3F Comparison using bar diagram between CI, II, HCR and MPI in the years 2019–2021.

Table 15.1. Status of Poverty and Food Insecurity of the Rural Sample Households in Odisha by Economic Characteristics.

	HCR	**Poverty Gap**	**Square Poverty Gap**
Status of Employment			
CL	53.3	9.2	1.86
RE	15.5	2.5	0.58
SE	24.7	2.4	0.39
Per capita Cultivable Land (in acre)			
0.01–0.50	43.4	5.2	0.96
0.51–1.00	3.4	0.2	0.01
More than 1 acre	7.8	0.1	0.0004
	Food Insecurity	**Food Insecurity Gap**	**Square Food Insecurity Gap**
Status of Employment			
CL	66.7	7.7	1.5
RE	20.7	4.3	1
SE	41.1	7.8	1.7
Per capita Cultivable Land (in acre)			
0.01–0.50	69	12.6	2.8
0.51–1.00	8.6	1.3	0.3
More than 1 acre	0	0	0

Source: Authors' calculation.

Table 15.1 indicates that the HCR, which signifies the proportion of households living below the poverty line, is highest among CL at 53.3%, followed by self-employed individuals at 15.5% and regular employees at 15.5%. Similarly, both the poverty gap and severity of poverty are highest among CL, with values of 9.2 and 1.86, respectively. However, despite the lower HCR percentage, the severity of poverty is comparatively higher among regular employees (0.58) than among the self-employed (0.39). This suggests that although a smaller proportion of regular employees fall below the poverty line, they experience deeper levels of poverty compared to self-employed households.

In the case of per capita landholding size, the HCR is observed to be highest for landholding sizes ranging from 0.01 to 0.50 acres, followed by those with more than 1 acre, and lowest for landholdings between 0.51 and 1 acre. Similarly, the poverty gap is most pronounced for landholdings of 0.01–0.50 acres, followed by those ranging from 0.51 to 1 acre, and least for landholdings exceeding 1 acre. The severity of poverty, represented by the squared poverty gap, is highest for landholdings between 0.01 and 0.50 acres, followed by those between 0.51 and 1 acre, and lowest for landholdings exceeding 1 acre, with values of 0.96, 0.01 and 0.0004, respectively.

The variation in poverty indicators across different categories of employment status and landholding sizes can be attributed to various socio-economic factors and structural inequalities within the population. For instance, CL often face irregular employment opportunities and lower wages, leading to higher levels of poverty incidence (HCR), poverty depth (poverty gap) and severity of poverty compared to regular employees and self-employed individuals. On the other hand, the severity of poverty may be higher for regular employees compared to self-employed individuals due to factors such as fixed incomes, dependence on specific industries or lack of diversification in income sources. Similarly, in the case of landholding sizes, smaller landholdings (0.01–0.50 acres) may indicate limited access to agricultural resources, insufficient productivity and higher vulnerability to economic shocks, contributing to higher levels of poverty. Conversely, larger landholdings (more than 1 acre) may provide greater opportunities for agricultural production, income generation and resilience against poverty. These disparities highlight the complex interplay of factors such as access to resources, employment opportunities, market conditions and government policies, which influence poverty dynamics within different segments of the population.

Food Insecurity Estimation for the Sample Households

The monthly proportion of per capita expenditure allocated to food serves as a key indicator in assessing household food security. Our study adopts this methodology to determine the food security status of the sampled households. Adopting the Rangarajan methodology and adjusting for inflation, we establish the food security threshold for rural Odisha at Rs 888.08, below which households are considered food insecure. Table 15.1 illustrates the food insecurity status of the sampled households, depicting the percentage of households falling below the food security line, the food insecurity gap and the squared food insecurity gap. Analysis reveals that 41% of households experience food insecurity, as their expenditure on food items falls below Rs 888.08. The food insecurity gap is calculated at 7.4, with a severity of 1.6.

Table 15.1 presents the food security status of sampled households categorized by employment status and per capita land holding size. Among the different employment categories, the highest proportion of food-insecure households is observed in the CL labour class (66.7%), followed by SE (41.1%) and RE (20.7%). However, when considering the food insecurity gap and severity, the self-employed class exhibits the highest values, followed by the CL and RE labour classes.

Similarly, households with per capita land holding sizes between 0.01 and 0.50 acres demonstrate the highest level of food insecurity, with 69% falling below the food security threshold. This group also exhibits the highest food insecurity gap of 12.6 and severity of 2.8. In contrast, 8.65% of households with per capita land holdings of 0.51–1 acre experience food insecurity, with a severity of 0.3. Notably, none of the households owning more than 1 acre of land are classified as food insecure.

Poverty and Food Insecurity Based on MPCE

Table 15.1 provides insights into both poverty and food security calculations based on the Monthly Per Capita Expenditure (MPCE) of households. It indicates that 20.5% of the sample households are classified as both poor and food insecure. Additionally, 5.6% are categorized as poor but food secure, implying that although they fall below the poverty line, they manage to allocate their expenditure above the food security threshold. A majority, comprising 53.4% of households, are classified as non-poor and food secure. However, a concerning 20.5% of households are identified as non-poor yet food insecure. These households earn above the poverty line, such as the threshold of Rs 1,558.04, but their food security remains compromised. This scenario poses a dilemma as these individuals are not eligible for government welfare schemes, despite their marginal income above the poverty line. Consequently, even a slight increase in income, such as Rs 1,560, does not alleviate their food insecurity, highlighting the limitations of existing policies and schemes in addressing their needs effectively.

The classification of households into various categories of poverty and food security is crucial for policymakers to design targeted interventions and allocate resources effectively. By analyzing the data presented in the table, policymakers can identify households that are most vulnerable and in need of immediate assistance. For instance, households classified as both poor and food insecure (20.5%) require comprehensive support to improve their economic situation and ensure access to an adequate diet. Similarly, households categorized as poor but food secure (5.6%) may benefit from targeted programmes aimed at enhancing their income-generating opportunities while ensuring sustained food security. Moreover, the presence of non-poor yet food-insecure households (20.5%) underscores the limitations of relying solely on prevailing poverty measures. These households, despite earning above the poverty line, struggle to meet their food needs, indicating broader systemic issues such as unequal access to resources, high food prices or inadequate social protection programmes. Addressing the challenges faced by these households requires a nuanced understanding of their specific circumstances and the implementation of policies that go beyond income support to tackle underlying structural factors contributing to food insecurity. Therefore, understanding the differences within different categories of households enables policymakers to tailor interventions effectively, ensuring that resources are directed towards those who need them the most and that no vulnerable group is left behind.

Regression Results

Descriptive Statistics

Table 15.2 displays the mean, standard deviation, minimum and maximum values of both dependent and independent variables considered in the study. The average

Table 15.2. Descriptive Statistics of Variables.

Notation	Specification of Variables	Mean	Std.	Min	Max
Dependent Variable					
	Both food secure and non-poor = 0 (53.4%)	0.9	1.2	0	3
	Food insecure but non-poor = 1(20.5%)				
	Food secure but poor = 2 (5.6%)				
	Food insecure and poor = 3 (20.5%)				
Independent Variables					
HAGE	Age of the household's head	53.2	7.3	35	65
SHAGE	Square of age of the household's head	2877.8	753.7	1,225	4,225
SE	Dummy variable takes the value 1 if any member of the household is self-employed and 0 otherwise	0.8	0.4	0	1
RE	Dummy variable takes the value 1 if any member of the household is regular employed and 0 otherwise	0.1	0.3	0	1
SOAGI	Share of agricultural income to total income	0.6	0.2	0.18	1
PCAGL	Per capita agricultural land	0.6	0.5	0.07	3
PCPDS	Per capita PDS consumption (in kg)	1.1	0.2	0.53	1.75

Source: Authors' calculation.

age of the household head is determined to be 53 years, as the head of the household typically makes decisions regarding major financial matters and expenditures. Consequently, the age of the household head is included as a variable to assess the household's financial status and food security, with the square of the age also examined to identify potential non-linear relationships with the dependent variable. Regarding employment, the study considers self-employment and regular employment, which are represented by dummy variables. The mean value for self-employment is 0.8, indicating that a significant portion of households are self-employed. In contrast, the average value for regular employment is only 0.1, suggesting that fewer household members hold regular employment positions. Furthermore, the analysis includes the share of agricultural income to total income, which averages 0.6, indicating that more than half of the household income comes from agriculture. Additionally, the average per capita agricultural land is measured at 0.6 acres, with a PDS consumption of 1.1 kg.

Regression Results of Multinomial Logit to Analysis to Examine Poverty and Food Security

Table 15.3 presents the estimated outcomes of the multinomial logit model with the marginal impacts of the explanatory variables. The overall model is a good fit as the *p*-value is equal to 0.000. A positive coefficient indicates that as the predictor's value rises, there is a probability of opting for alternative categories instead of the base category. The analysis of the outcomes for the three alternative categories – Non-poor but Food Insecure, Poor but Food Secure and both Food Insecure and Poor – are explained below:

- Non-poor but Food Insecure
  The variables SOAGI, PCAGL and PCPDS exhibit significant negative effects on the category of households classified as non-poor but food-insecure. A higher proportion of agricultural income to total income, greater per capita agricultural land and increased per capita PDS consumption are less likely to be non-poor but food insecure relative to non-poor and food secure with a marginal effect of −0.093, −0.829 and −0.491, respectively. This is due to the fact that the majority of the sample households are agriculture-based, most of which are small and marginal farmers possessing limited per capita land holdings; they predominantly depend on income generated from agricultural activities and the food produced on their own farms to meet their food expenditure needs. Additionally, the PDS significantly contributes to improving food security in rural regions by providing access to subsidized food items. This is reflected in the negative marginal effect observed in the category of non-poor but food-insecure households.
- Poor but Food Secure
  The variable RE only shows a significant negative effect on the category of households classified as poor but food secure, with a marginal effect of −0.05. The negative effect of the RE variable on the category of poor but food-secure

Table 15.3. Regression Results.

Number of Observations = 100.
LR Chi²(21) = 90; Prob > Chi² = 0.0000
Log Likelihood = −68.4; Pseudo R^2 = 0.397
0 = Base Category (Food Secure and Non-poor)

Variables	1 = Food Insecure but Non-poor				2 = Food Secure but Poor				3 = Both Food Insecure and Poor			
	Coef.	Robust Std. Err.	*z*	*dy/dx*	Coef.	Robust Std. Err	*z*	*dy/dx*	Coef.	Robust Std. Err.	*z*	*dy/dx*
HAGE	−0.35	0.60	−0.58	−0.057	−0.56	0.64	−0.87	−0.04	−0.56	0.76	−0.74	−0.001
SHAGE	0.004	0.006	0.67	0.0006	0.005	0.006	0.79	0.0003	0.006	0.007	0.75	0.00001
SE	−0.47	1.16	−0.4	−0.067	−0.10	1.42	−0.07	0.0019	−3.58	1.76	−2.03**	−0.029
RE	−0.56	1.71	−0.33	−0.078	−0.61	2.01	−1.62*	−0.05	−3.82	2.44	−1.56*	−0.003
SOAGI	−2.21	1.22	−1.81*	−0.093	−2.42	2.60	−0.93	−0.14	4.55	2.75	1.66*	0.009
PCAGL	−5.82	1.71	−3.4***	−0.829	−2.21	1.83	−1.21	−0.06	−20.33	5.06	−4.02***	−0.035
PCPDS	−3.16	1.82	−1.74*	−0.491	−2.08	2.75	−0.75	0.16	−3.64	2.10	−1.74*	−0.006
Intercept	−0.70	15.03	−0.05		13.10	15.76	0.83		−1.63	19.13	−0.09	

Source: Author's calculation.

Note: *t*-test is statistically significant where *** denotes 1% level of significance, ** denotes 5% level of significance and * denotes 10% level of significance.

households implies that an increase in the proportion of households engaged in regular employment leads to a decrease in the likelihood of being classified as poor but food secure. This is because regular employment provides a stable and consistent source of income, which can help lift households out of poverty and ensure their food security.

- Poor and Food Insecure
  The variables SE, RE, PCAGL and PCPDS significantly negatively affect the category of poor and food-insecure households, with marginal effects of −0.029, −0.003, −0.035 and −0.006, respectively. This means that as the self-employed and regularly employed class of employment, per capita agricultural land holding and per capita PDS consumption increase, the households are more likely to be non-poor and food secure over poor and food insecure. This explains that better income opportunities will provide a more stable source of income and benefits, reducing household's vulnerability to poverty and food insecurity. Moreover, households with larger landholdings have greater access to resources for food production or income generation from agricultural activities, reducing their dependence on external sources for food and providing better access to affordable food through PDS will contribute to food security and poverty alleviation. On the other hand, SOAGI shows a significant positive effect on the category of households classified as poor and food insecure, with marginal effects of 1.66. This implies that as the share of agriculture income to the total income increases, the households are more likely to be poor and food insecure. This positive effect can be attributed to the reliance on agriculture as the main source of income. When a higher proportion of income comes from agriculture, where the majority are small and marginal farmers, it indicates a dependence on an inherently volatile and uncertain source of income. Agricultural income is often subject to fluctuations due to factors such as weather conditions, market prices and crop yields. Consequently, households reliant on agriculture may face greater economic vulnerability.

Conclusion and Policy Recommendations

The study provides a comprehensive analysis of poverty and food security dynamics based on MPCE in rural Odisha, drawing on a robust analysis of household-level data. Through a multi-stage random sampling method, 100 households were interviewed from two villages in the Balasore district. Using the m-logit model, the study indicates interplay of socio-economic factors that underpin poverty and food insecurity, shedding light on both the progress made and the persistent disparities that exist. Significant variables such as the share of agricultural income to total income, per capita agricultural landholding and per capita PDS consumption impact household poverty and food security outcomes. Notably, 20.5% of households classified as non-poor but food insecure highlight a vulnerable segment often overlooked by government programmes. Regular and stable employment, particularly in agriculture, emerges as a key factor in mitigating poverty and ensuring food security.

The findings of this study carry several important policy implications for addressing poverty and food insecurity in rural Odisha. As the results showed location-specific clustering of these issues, there is a need for targeted interventions that address the unique needs and vulnerabilities of different segments of the population based on their location and socio-economic characteristics. Since there is a prevailing significant contribution of agriculture to household incomes, there is a need for increased support for smallholder farmers, including access to credit, technology and market linkages. Investments in irrigation infrastructure and agricultural extension services through an inclusive integrated assessment approach to promote sustainable farming practices, including nutrition-sensitive agriculture (NSA), and improve access to markets and technology are critical components of building long-term food security (Merritt et al., 2022). It can also help to improve productivity and resilience in the face of climate variability. Efforts to strengthen social protection programmes, such as the PDS, can play a crucial role in enhancing food security among vulnerable households. This includes expanding the coverage of PDS to reach underserved areas and ensuring the availability of nutritious food items at affordable prices.

References

Aboaba, K., Fadiji, D. M., & Hussayn, J. A. (2020). Determinants of food security among rural households in Nigeria: USDA food insecurity experience based measurement approach. *Journal of Agribusiness and Rural Development*, *56*(2), 113–124.

Ali, A., & Erenstein, O. (2017). Assessing farmer use of climate change adaptation practices and impacts on food security and poverty in Pakistan. *Climate Risk Management*, *16*, 183–194.

Ballard, T. J., Kepple, A. W., Cafiero, C., & Schmidhuber, J. (2014). Better measurement of food insecurity in the context of enhancing nutrition. *Ernahrungs Umschau*, *61*(2), 38–41.

Banerjee, A. (2016). Challenges to food security in South Asia. Working Papers id: 11275. *eSocialSciences*. https://ideas.repec.org/p/ess/wpaper/id11275.html

Baral, N., Paria, B., Behera, B., & Mishra, P. (2021). Household behaviour and nutrition-sensitive agricultural practices: Experiences of smallholder farmers in Northern West Bengal, India. *World Development Perspectives*, *21*, 100296.

Bay, U. (2022). Radical approaches to anti-poverty strategies. In S. S. Webb (Eds), *The Routledge handbook of international critical social work: New perspectives and agendas* (pp. 302–314). Taylor & Francis.

Castell, G. S., Rodrigo, C. P., de la Cruz, J. N., & Bartrina, J. A. (2015). Household food insecurity access scale (HFIAS). *Nutricion Hospitalaria*, *31*(3), 272–278.

Chauhan, R. K., Mohanty, S. K., Subramanian, S. V., Parida, J. K., & Padhi, B. (2016). Regional estimates of poverty and inequality in India, 1993–2012. *Social Indicators Research*, *127*, 1249–1296.

Clay, E. (2002). Food security: Concept and measurement. Paper for FAO Expert Consultation on *Trade and food security: Conceptualising the linkages* (pp. 25–34).

Das, P., Ghosh, S., & Paria, B. (2022). Multidimensional poverty in India: A study on regional disparities. *Geojournal, 87*(5), 3987–4006.

FAO, IFAD, UNICEF, WFP and WHO. (2021). *The State of Food Security and Nutrition in the World 2021. Transforming food systems for food security, improved nutrition and affordable healthy diets for all.* FAO. https://doi.org/10.4060/cb4474en

FAO. (2006, June). [Policy Brief] *Food security*. Agriculture and Development Economics Division (ESA). https://www.fao.org/fileadmin/templates/faoitaly/documents/pdf/pdf_Food_Security_Cocept_Note.pdf

Frongillo, Jr, E. A. (1999). Validation of measures of food insecurity and hunger. *The Journal of Nutrition, 129*(2), S506–S509.

Gweshengwe, B., & Hassan, N. H. (2020). Defining the characteristics of poverty and their implications for poverty analysis. *Cogent Social Sciences, 6*(1), 1768669.

Heidhues, F., Atsain, A., Nyangito, H., Padilla, M., Ghersi, G., & Vallée, L. (2004). *Development strategies and food and nutrition security in Africa: An Assessment*, (Vol. 38). International Food Policy Research Institute.

Hulme, D., & McKay, A. (2013). *Identifying and measuring chronic poverty: Beyond monetary measures?* Palgrave Macmillan.

Irfeey, A. M. M., Najim, M. M. M., Alotaibi, B. A., & Traore, A. (2023). Groundwater pollution impact on food security. *Sustainability, 15*, 4202. https://doi.org/10.3390/su15054202

Isazade, V., Qasimi, A. B., Dong, P., Kaplan, G., & Isazade, E. (2023). Integration of Moran's I, geographically weighted regression (GWR), and ordinary least square (OLS) models in spatiotemporal modeling of COVID-19 outbreak in Qom and Mazandaran provinces, Iran. *Modeling Earth Systems and Environment, 9*(4), 3923–3937.

Kendall, A., Olson, C. M., & Frongillo Jr, E. A. (1995). Validation of the Radimer/Cornell measures of hunger and food insecurity. *The Journal of Nutrition, 125*(11), 2793–2801.

Martin, W. (2010). *Food security and poverty—A precarious balance*. World Bank Blogs: Let's Talk Development. https://blogs.worldbank.org/developmenttalk/food-security-and-poverty-a-precarious-balance

McKay, F. H., Sims, A., & van der Pligt, P. (2023). Measuring food insecurity in India: A systematic review of the current evidence. *Current Nutrition Reports, 12*(2), 358–367.

Merritt, W. S., Hamilton, S. H., Bagchi, N. S., Baral, N., Carter, L., Chakraborty, A., & Williams, L. J. (2022). Reflecting on an integrated approach to understanding pathways for socially inclusive agricultural intensification. *Journal of Development Studies, 58*(8), 1569–1587.

Narain, S., Singh, S. K., & Singh, L. (2016). Impact of national food security mission–pulse on chickpea productivity in Hamirpur district of Uttar Pradesh. *Indian Research Journal of Extension Education, 14*(3), 24–27.

Narayan, D., Chambers, R., Shah, M. K., & Petesch, P. (2000). *Voices of the poor: Crying out for change*. Oxford University Press for the World Bank.

Oluoko-Odingo, A. A. (2011). Vulnerability and adaptation to food insecurity and poverty in Kenya. *Annals of the Association of American Geographers, 101*(1), 1–20.

Patnaik, U. (2005). Theorizing food security and poverty in the era of economic reforms. *Social Scientist*, 50-81.
Patra, R. (2014). Agricultural development in Odisha: Are the disparities growing?. *The International Journal of Food and Agricultural Economics*, *2*(3), 129–144.
Pérez-Escamilla, R. (2017). Food security and the 2015–2030 sustainable development goals: From human to planetary health: Perspectives and opinions. *Current Developments in Nutrition*, *1*(7), e000513.
Pradhan, R. K. (2017). *Regional disparities in Odisha: Factors responsible for the persisting regional imbalances in Odisha*. http://doi.org/10.2139/ssrn.2934740
Radimer, K. L., Olson, C. M., & Campbell, C. C. (1990). Development of indicators to assess hunger. *The Journal of Nutrition*, *120*, 1544–1548.
Raphael, D. (2011). A discourse analysis of the social determinants of health. *Critical Public Health*, *21*(2), 221–236.
Saxena, N. C. (2018). Hunger, under-nutrition and food security in India. In A. Mehta, S. Bhide, A. Kumar, A. Shah, (Eds), *Poverty, chronic poverty and poverty dynamics*. Springer. https://doi.org/10.1007/978-981-13-0677-8_4
Sen, A. (1982). *Poverty and famines: An essay on entitlement and deprivation*. Oxford University Press.
Swedish International Development Cooperation Agency (SIDA). (2015). [Portfolio Overview] Top of Form *Food security and agriculture: Secure food for a growing world and end hunger*.
World Food Conference (1974: Rome, Italy). (1975). *Report of the World Food Conference, Rome, 5–16 November, 1974*. United Nations.
World Food Summit. (1996). *Rome declaration on world food security and world food summit plan of action*. FAO.

Index

Printed and bound by CPI Group (UK) Ltd, Croydon, CR0 4YY
15/09/2026
14962489-0005